Incident Commander for Ground Search and Rescue

Student Manual

by
Robert J. Koester

dbS Published by
dbS Productions

Copy Editor: Mary Hardy

Published by dbS Productions
P.O. Box 1894
University Station
Charlottesville, Virginia 22903
(800) 745-1581

dbS

Copies of the Task Assignment Form used by permission of the Virginia Search and Rescue Council
Fig 3.19-21 adapted from *Weathering the Wilderness* used by permission from Sierra Club
Clipart provided by license from Corel Corporation.

Printed in the United States

10 9 8 7 6

Publisher's - Cataloging in Publication

Koester, Robert James, 1962-
    Incident Commander for Ground Search and Rescue: Student Manual/
 by Robert J. Koester. -
230p.; 21.5cm
Includes Glossary Appendices
ISBN 978-1-879471-57-3
Library of Congress Numbering: 97-68957

CONTRIBUTORS

David Carter, MSW, Incident Commander and Chairman, Appalachian Search & Rescue Conference; Chief Instructor, Virginia Department of Emergency Services Ground Search & Rescue Training Program; President, Search and Rescue Training Association.

William H. Dixon, MS, Incident Commander and Vice-Chairman, Appalachian Search & Rescue Conference; Technical Instructor, Virginia Department of Emergency Services.

Greg Fuller, Tracker, Appalachian Search & Rescue Conference; Director, Search and Rescue Tracking Institute; Tracking Program Director, Virginia Department of Emergency Services.

Gilbert Gray, M.ED., Group 2 Commander, Incident Commander/Mission Coordinator, Virginia Wing Civil Air Patrol.

Robert J. Koester, MS, Incident Commander, Appalachian Search & Rescue Conference; President, Virginia Search & Rescue Council; Technical Instructor and Principle Investigator, Virginia Department of Emergency Service; Research Scientist, dbS Productions.

Mark Pennington, Incident Commander, Appalachian Search & Rescue Conference; Search and Rescue Duty Officer, SAR Instructor, Virginia Department of Emergency Service; Search Team Manager/Agency Liaison, FEMA USAR Task Force VA-2.

Introduction

As a comparative youngster to the field of search and rescue I have been fortunate to witness many changes and innovations in the field. Search management courses are well developed and available, Incident Command System (ICS) is becoming universal, lost person behavior is becoming more scientific, and the list goes on. However, the practice of SAR is still a mixture of art and science. New research continues to change and mold the art of finding a missing person. The nasty word of *standards* has crept into the often chaotic world of SAR and the practice has been forced to change for the better. Yet the aged grumpy team member who simply points at the map and always locates the missing subject is still invaluable.

Selection of an Incident Commander (IC) may range from whoever arrives on scene first to a system of standards and qualifications. This student manual and its complimentary course has arisen out of an identified need to provide high quality and advanced training to new Incident Commanders. While the manual may stand alone as a reference manual, it is primarily intended to be an integral part of a training course. As the Appalachian Search & Rescue Conference (ASRC) grew from two teams to seven teams in three states with 400 members, the need for trained Incident Commanders increased. The Virginia Department of Emergency Services continued to push for higher performance through training programs and statewide training standards. Initially, *Managing the Search Function* (MSF) or *Managing Search Operations* (MSO) was the accepted standard for the "Incident Commander." With both the state and the ASRC offering numerous courses each year, it became the entry-level course for new staff members. Then, Incident Commanders were required to demonstrate proficiency while serving as staff members before becoming qualified. The effort of continual improvement has pushed the standards even higher. The Virginia Department of Emergency Services began offering a 40-hour course called *Practical Search Operations* that built upon a MSO or MSF training. This course provided new staff members and graduates of MSO/MSF with practical knowledge of how to run a search and manage the information flow (paperwork). In 1994 the Virginia Department of Emergency Services then developed Incident Commander standards. In the standards, a finishing course was required. The standards also reflected the realization that Incident Commanders must be well schooled in both operations and plans, since they often must fulfill or teach these to general staff members during smaller operations. This textbook arose to fulfill the need of training new Ground Incident Commanders.

This document is based largely upon the personal experience of the author during more than 100 operational periods as an Incident Commander. Extensive credit must be given to the *Incident Command for Search and Rescue, Standard Operating Procedures* manual developed by David Carter and *Managing Search Operations* by Rick LaValla and Skip Stoffel. I must also acknowledge several forms and procedures developed by the Virginia Search and Rescue Council, the Virginia Department of Emergency Services, and the ASRC. Finally, the book would not exist without long discussions held with many ICs and staff members of the ASRC and NPS.

Table of Contents

Instructional Units

Appendices

Rationale

The position of Incident Commander (IC) is required on every search and often determines the success or failure of the search effort. The Incident Commander determines the overall direction of the search through leadership, selection of staff, managing external influences, resource ordering, organization of the search structure, information management, and establishment of mission objectives. Incident Commanders are typically selected due to field and staff experience. However, Incident Commanders require training and a set of skills unique to the position.

Incident Commander candidates need a training course that helps develop IC skills. The opportunity for feedback and refinement can be gained through practical exercises, demonstration of skills, and scenario playing. This course offers the IC candidate specific knowledge necessary to fulfill the role, practical exercises to refine important skills, and search simulations that allow experienced instructors an opportunity to provide feedback to the student.

Scope of Purpose

The purpose of the *Incident Commander for Ground Search and Rescue* (ICG) course is to help prepare the student to function as a Type III Incident Commander. The course will allow the student to obtain didactic knowledge required to fulfill the knowledge objectives of The Commonwealth of Virginia IC-ground standards. The course requires utilization of previous SAR knowledge and training through the use of several practical exercises. Graduates of the class that meet additional experience requirements are expected to be able to function as staff members on type II incidents (multiple SAR organizations present) and fulfill the Incident Commander function on type III incidents (single SAR agency present). Establishment of common standards for different levels of Incident Commanders will allow better interface and understanding across SAR teams.

Course Description

Incident Commander For Ground Search and Rescue (ICG)   (24 Hours)
ICG is an advanced management course for experienced SAR personnel. It goes beyond theoretical MSO training and prepares the student to manage complex type II (multi-agency) searches as General Staff members or as a type III Incident Commander. Subjects include ICS, SAR operations and tactics, resource management, SAR legal documents, IC roles and responsibilities, ground aspects of aircraft searches, command staff, and plans. The course makes extensive use of map problems, practical exercises, simulations, handouts, and lectures. MSO or equivalent, a working knowledge of ICS, and SAR experience are required prerequisites.

Overall Course Objectives

- Describe methods of personal evaluation of job performance.
- Develop a personal objective for all ground SAR incidents.
- Describe how to supervise, train, and perform all staff functions common to a ground search and rescue incident.
- Provide a working knowledge of the NIIMS Incident Command System.
- Demonstrate an understanding of SAR resources.
- Demonstrate the ability to establish and manage SAR information systems.
- Demonstrate the ability to brief and debrief teams and staff members.
- Describe all SAR resources, including planning and operational considerations.
- Describe legal issues and documents relevant to SAR.
- Demonstrate proper tasking of ground resources used in a missing aircraftincident including use of investigation and National Track Analysis Program (NTAP) data.
- Describe Incident Commander roles and responsibilities.
- Describe how to properly handle external influences.
- Demonstrate how to give an effective press interview.
- Develop and evaluate a complete Incident Action Plan (IAP).

Participant Qualifications

Students in the Incident Commander for Ground Search and Rescue course must be experienced search and rescue providers. The student must have demonstrated field leadership on searches through serving as a leader of a field task. Ideally students will be certified as Field Team Leaders, Operational Dog Handlers, Mounted Horse Riders, or Trackers capable of leading a task. In addition, the student must have a theoretical background in search management by successful completion of a Managing Search Operations (MSO) or Managing the Search Function (MSF) class. Additional staff knowledge must be obtained through Practical Search Operations (PSO) or equivalent experience. Students who have functioned in a base position during an incident will be at a strong advantage when taking the class. The student should also have a working knowledge of the Incident Command System (ICS). This can be gained through participation on search incidents where ICS was used extensively or through ICS course work. The student should posses an openness to learning different techniques and a willingness to receive instructor feedback on his or her areas of difficulty.

Students should be recruited who have the personality characteristics that predict success as a leader. Possession of knowledge does not predict success as an Incident Commander or Staff Member. The ability to work in a team environment under stress is a better predictor of success in SAR management.

Instructor Qualifications

Instructors in the *Incident Commander for Ground Search and Rescue* course must be experienced type II or type I ground SAR Incident Commanders. Since the course is intended to be highly interactive, it is important to have instructors who are skilled in group processes and conflict resolution (typical attributes of type II Incident Commanders). Instructors should be experienced and dynamic teachers skilled in assisting students to develop independent thinking. Associate instructors may be used to address areas of specific knowledge and training. Instructors with strong opinions and philosophical or professional positions that only one method of management is possible could have a negative impact on the achievement of the course objectives. Attention should be given to providing a creative learning environment that will enable participants to determine what will work best under different situations.

Sample Course Outline

Day One	Day Two	Day Three
	Unit 4: SAR Operations Unit 5: SAR Resources Unit 6: Legal and Documents	Unit 9: Command Staff Unit 10: Plans Unit 11: Logistics Unit 12: Finance
Lunch		
	Unit 7: Ground OPS for Missing Aircraft Unit 8: IC Roles and Responsibilities	Testing Unit 14: Course Evaluation.
Dinner		
Unit 1: Introduction Unit 2: ICS Implementation Unit 3: SAR Tactics	Unit 13: Simulations	

Unit Titles and Scopes

Unit 1: Introduction   Time: 0.5 hours
Welcome; staff and participant introductions; administrative information; course objectives; search philosophies; role of staff and Incident Commander; expectations and the customer; course caveats.

Unit 2: Incident Command System Implementation   Time: 1.0 hour
ICS general concepts; different ICS structures used in SAR; unified Command; scenario-based problem; creating command structures; scenario-based problem of operating in an Unified Command.

Unit 3: Operational Tactics   Time: 2.0 hours
Detailed tactical aspects of deploying air and ground SAR resources; scenario-based problems of resource deployment; evaluation of written tasks.

Unit 4: Operations   Time: 1.0 hour
IC overview of operations and information flow; the role of the operations officer and staff; operational information management and resource tasking; urban search factors; scenario-based problem of managing information; task assignment form evaluations.

Unit 5: SAR Resources   Time: 1.5 hour
Planning factors to consider before placing a resource order of federal, state, SAR, and auxiliary resources common to SAR incidents.

Unit 6: Legal and Review of Applicable SAR Documents   Time: 1.0 hour
SAR legal issues, legal documents, SAR Standard Operating Procedures, SAR documents, scenario-based problem solving of legal problems, scenario-based resource order.

Unit 7: Ground Operation for Missing Aircraft   Time: 1.5 hours
Flight planning, navigation, and aircraft system; crash statistics and behavior; air search techniques; ground operations; ground investigations; clue management; NTAP; scenario-based aircraft search.

Unit 8: Incident Commander Roles and Responsibilities   Time: 1.0 hour

IC operational duties; initial role; second operational period; managing external influences; staff management techniques; common IC mistakes; find/rescue issues; mission suspension process; mission review process.

Unit 9: Command Staff   Time: 0.5 hour

Roles of level III and level II Incident Commanders; use of information, safety, and liaison officers; scenario-based media interview.

Unit 10: Plans   Time: 1.5 hours

IC oversight of planning and the Incident Action Plan; use of the combined plans and operations (PLOPS) function; planning section chief and staff roles; planning; maps and process; development of the IAP; investigations; scenario-based IAP writing and evaluation.

Unit 11: Logistics   Time: 0.5 hours

IC oversight of logistics; logistics section chief and staff roles. Overview of the logistics process.

Unit 12: Finance/Administration   Time: 0.5 hours

IC oversight of finance; finance section chief and staff roles; accident compensation process.

Unit 13: Simulated Exercise   Time: 3.0 hours

Scenario-based simulated search requiring total integration of skills and functions.

Unit 14: Course Evaluation   Time: 0.5 hour

Course conclusion; participant evaluation of course effectiveness and instructors.

Unit 15: Practical Exercises   Time: dependent

List of practical exercises used throughout the course.

Unit Objectives

Unit 1: Introduction

Understand the five major Incident Commander (IC) roles:

- ☐ Be able to perform, evaluate, and teach every staff position.
- ☐ Learn how to handle external influences
- ☐ Establish, monitor, and control staff structure and information flow.
- ☐ Maintain safety while meeting mission objectives.
- ☐ Provide leadership.

Unit 2: Incident Command System Implementation

- ☐ Demonstrate an understanding of the NIIMS Incident Command System as it applies to SAR and how the system can be adapted to any size incident.
- ☐ Demonstrate the ability to work within an unified command system with both SAR trained and untrained incident commanders.
- ☐ Describe different structures for using the "PLOPS" function.
- ☐ Describe factors that influence the structure of ICS.

Unit 3: Operational Tactics

- ☐ Demonstrate an understanding of SAR resources including appropriate and inappropriate uses of, ground teams, air-scent dogs, trailing/tracking dogs, mantrackers, mounted horses, helicopters, fixed wings, technical teams, ATVs, bicycle teams, and kayak teams.

Unit 4: Operations

- ☐ Demonstrate the ability to develop and manage a staff and describe when, where, and why various functions should be assigned to which staff positions, including operations and clue analysis.
- ☐ Describe the internal staff information flow system, including verbal, written, and electronic communications, which is required in order to insure that information is properly collected, evaluated, disseminated, utilized, and stored throughout the incident.
- ☐ Describe the differences in deployment of resources in urban, suburban, rural, and wilderness searches.
- ☐ Demonstrate the ability to complete all necessary operational mission documentation.

Unit 5: SAR Resources

- ☐ Demonstrate a complete understanding of SAR resources, including important planning considerations.
- ☐ Demonstrate an understanding of non-SAR resources, including how they are obtained and their potential function in a SAR incident.
- ☐ Describe how to effectively and efficiently use non-SAR resources, which may offer help at all types of searches.
- ☐ Describe how to place a resource order.

Unit 6: Legal and Review of Applicable Documents

☐ Demonstrate an understanding of the laws, policies, procedures, operating instructions, memorandums and agreements which govern SAR operations.

☐ Demonstrate an understanding of certain legal issues related to SAR including, temporary restraining order, trespassing, confidentiality, criminal investigations, management of deceased subjects, restricted airspace, restricting access to various areas, site security and surveillance, maintaining the chain of evidence, use of minors, liability for supplies and equipment, use of SAR personnel for apprehension of criminals and crime scene investigation, discovery of non-incident related illegal activities.

☐ Describe when and how to contact DES and what type of incident information DES requires.

Unit 7: Ground Operations for Missing Aircraft

☐ Briefly review plans and agreements governing air SAR.

☐ Understand basic tools used in flight operations, including flight plans, aeronautical charts, and airport directories.

☐ Discuss aircraft communications and navigational systems.

☐ Review types of SAR aircraft, including their use and limitations.

☐ Discuss crash statistics and behavior.

☐ Discuss air search planning and techniques.

☐ Discuss ground operational planning and techniques.

☐ Discuss scene control.

☐ Discuss elements of clue management.

Unit 8: IC Roles and Responsibilities

☐ Describe the roles and responsibilities an IC may perform upon being assigned as an IC, enroute to the incident, upon initial arrival, after initial tasks, during the second operational period, and at the end of a search.

☐ Describe important considerations in selecting a replacement IC and the process of IC transition.

☐ Describe the role of the IC in relation to the Legal Responsible Agent/Agency Administrator (RA/AA) when the RA is uncooperative and when the mission involves or expands into other jurisdictions.

☐ Describe the role of the IC in relation to the various resources that may participate in a search mission in the following situations: when the IC has overall responsibility for all resources present, when there are resources present that may not be willing to cooperate.

☐ Demonstrate the ability to communicate with staff by means of briefings, meetings, and written communications.

☐ Describe common mistakes inexperienced Incident Commanders make and how to avoid them.

☐ Identify outside influence problems common to search missions such as family, media, politicians, and psychics. Describe solutions and justify them.

☐ Describe the process used when deciding to suspend a mission.

☐ Explain the IC's role after the subject or target has been located.

Unit 9: Command Staff

☐ Describe the roles and responsibilities of the Information Officer, Safety Officer, Liaison Officer, and the Family Liaison Officer.

☐ Describe methods for giving an effective media interview.

☐ Identify potential safety issues and describe how they can be mitigated if possible.

☐ Describe when risk factors outweigh the need to continue operations.

☐ Describe the common signs of incident stress and define the criteria for recommending a critical incident stress debriefing.

☐ Describe methods for keeping the family informed.

Unit 10: Plans

☐ Understand role of PSC and investigations.

☐ Understand function of PSC during initial shift and its relationship to OPS.

☐ Demonstrate the ability to determine search areas using theoretical, statistical, deductive, and subjective process.

☐ Demonstrate the ability to determine planning segments and search sectors and perform Mattsons.

☐ Demonstrate the ability to make an IAP including a 202, 203, 204, 205, demobilization, SISs, rescue/evac, press, and base map.

☐ Demonstrate the ability to properly document a search.

Unit 11: Logistics

☐ Describe the role and responsibilities of the Logistics Section Chief and staff.

Unit 12: Finance/Administration

☐ Describe the role and responsibilities of the Finance Section Chief, Time Unit Leader, Procurement Unit Leader, and Compensation Injury Specialist.

☐ List situations that would require activation of the elements found within the Finance/Administration Section.

☐ Describe methods the IC may use to contain costs.

Unit 13: Simulations

☐ Demonstrate the ability to effectively manage staff, external influences, and information flow; generate tasks; and efficiently deploy field resources given a search scenario, staff, and field resources.

Unit 14: Course Evaluation

☐ Obtain feedback from class participants on the presentation, materials used, and usefulness of each instructional unit in order to continually improve the course.

Unit 15: Practical Exercises

☐ Provide a list of practical exercises used throughout the course.

UNIT 1: INTRODUCTION

I. Introduction

A. Administration
  1. Welcome students
  2. Introduce students/instructors
  3. Course Overview
     a. Provide information through lectures and materials.
     b. Encourage students to think through role playing, practical exercises, map problems, and simulations.
  4. Review objectives
  5. Student expectations
  6. Pass out handouts
  7. Pass out schedule
  8. Ensure sign in
  9. Review facility
  10. Course assumptions (prerequisites)
  11. The Incident Commander (IC) Role
     a. Nothing really prepares you for the IC position.
     b. IC is completely different than any other Incident Command System (ICS) position.
     c. Each search is unique and needs different strategies and structures.
     d. A true IC exudes leadership, personality, experience, and knowledge.
     e. An IC must have a thirst for additional knowledge.

> **Objectives:**
> Understand the five major Incident Commander (IC) roles:
> ☐ Be able to perform, evaluate, and teach every staff position common to ground search & rescue.
> ☐ Learn how to handle external influences.
> ☐ Establish, monitor, and control staff structure and information flow.
> ☐ Maintain safety while meeting mission objectives.
> ☐ Provide leadership and vision.

> **Nobody ever said that staff work was easy**

B. Operational Philosophy
  1. Assists in evaluation of all decisions
  2. Similar to a mission statement
  3. Every IC needs to develop a clear idea of what they want to accomplish.
  4. Examples
     a. The subject would approve of this decision.
     b. Koester's
        (1) Safely
        (2) Locate, access, stabilize, and transport the subject
        (3) in the shortest possible time frame
        (4) with the most efficient type and number of resources
        (5) while following any applicable laws, rules, and regulations.

C. The IC Qualification System
1. Type IV: First Responder- Initial attack or first response to an incident
   a. Usually law enforcement or local fire-rescue
   b. Typically not SAR trained or experienced
   c. Ideally SAR first responder course and preservation of the PLS
   d. IC role: IC is a "hands on" leader, IC typically performs all the functions of OPS, LSC, PSC, and FSC.
   e. Duration: Short
   f. Examples: Minor carry-out, short rescue such as an ankle injury

2. Type III: Single Agency (SAR teams)
   a. First SAR team on scene
   b. A type III IC would typically function as general staff (Operations, Plans Section Chief, Logistics Section chief) on multi-agency searches.
   c. IC role: IC walks the line between manager and a doer. May have functional support positions filled.
   d. Duration: Does not usually go into another operation period /12 hours.
   e. Examples: Activation of a SAR team; evidence search; water search; most technical rescues that can be handled by a single SAR resource.

3. Type II: Multi-agency (multi-group/division).
   a. Required when multiple types or number of SAR teams arrive.
   b. Sometimes referred to as State IC.
   c. IC role: IC spends all their time as a manager.
   d. Duration: Operational periods (12 hours) highly likely unless a find is made.
   e. Teams: The National Park Service has many All-risk regional type II teams. A team consists of an IC, staff, and equipment. Teams have sufficient depth to account for typical unavailability.
   f. Examples: Most state run searches that involve 1-2 dog groups, 1-2 ground SAR groups, and local resources. Lengthy searches with extensive paperwork requirements.

4. Type I: Multi-Branch
   a. Seldom required except for the largest, most complex searches that require multiple branches.
   b. An IC at this level can typically function as an Area Command Authority IC, who allocates resources when multiple missions occur.
   c. IC role: IC and some elements of the general staff serve as managers.
   d. Duration: Operational periods (12 hours) are guaranteed. Demobilization after a find will generally take an additional operational period.
   e. Teams: The National Park Service has one All-risk type I team.
   f. Examples: Manasas "Six Friends," Hurricane Andrew.

Incident Commander for Ground SAR

ICS Designator	ICS Level Description	Required Courses/Level	Practical Experience
Type IV IC (GSAR)	First Responder	SAR First Responder	None
Type III IC (GSAR)	Single SAR agency	MSO/MSF PSO ICGSAR	SAR agency member, Leader on 6 ground field tasks, Experience in 2 base positions, Letter of support from group, State approval.
Type II IC (GSAR)	Multi-SAR Agencies Multi-group/division	Type III IC	Meet experience standard, Type III IC experience/Type II IC simulation, support from group, Type II IC sponsor, Letters of support from ICs, State approval
Type I IC (GSAR)	Multi-Branch	Type II IC PIO	10 IC experiences on type II searches, Letter of support from group, Type I IC sponsor, State approval

Incident Staff for Ground SAR

ICS Designator	ICS Level Description	Required Courses/Level	Practical Experience
Type IV IS (GSAR)	First Responder	SAR First Responder	None
Type III IS (GSAR)	Single SAR agency	FTL/ODH MSO/MSF	SAR agency member, Leader on 2 ground field tasks, letter of support from group, state approval
Type II IS (GSAR)	Multi-SAR agencies Multi-group/division	Type III IS PSO	Serve in base on two searches, letter of support from group, state approval
Type I IS (GSAR)	Multi-Branch	Type II IC	State approval

Incident Management Teams for Ground SAR

ICS Designator	ICS Level Description	Required Personnel	Required Equipment
Type IV (GSAR) Management Team	First Responder	2 SAR First Responder	Individual-level operations kit
Type III (GSAR) Management Team	Single SAR agency	2 Type III IC 4 Type III IS	Team-level operations kit
Type II (GSAR) Management Team	Multi-SAR Agencies Multi-group/division	2 Type II IC 8 Type II IS	2 Team-level operations kits 1 base radio package, 4 handheld radios, 1 cellular/satellite phone, 1 portable copier, 3 map boards with set of 7.5 minute topos for coverage area.
Type I (GSAR) Management Team	Multi-Branch	2 Type I IC 8 Type I IS 8 Type II IS	4 Team-level operations kits 3 base radio packages, 12 handheld radios, 1 cellular/satellite phone, 3 portable copiers, 6 map boards with set of 7.5 minute maps

D. Caveats
1. Flexibility
2. Innovation
3. Different philosophies
4. Learn from experience (admit mistakes)
5. Don't think too much

II. **SAR Fundamentals**
  A. Central Themes
    1. GET THE TEAMS OUT!!
       a. Quick response requires active pressure from IC
       b. First team out most important
       c. Why are staff slow?
          (1) Desire to collect more information
          (2) Desire to talk about different task ideas
          (3) Unsure what to do next because tasking system not set up
          (4) Not sure what needs to be done next
    2. Constant pressure required at each stage

> **Generally, you don't find the subject when all the teams are in base.**

3. Reflex tasks
   a. Definition
   b. Examples
4. Resource mixing
   a. Definition
   b. Purpose
   c. Examples
5. Always know where your teams are located
6. NEVER forget the statistical search
7. NEVER forget the basics
   a. The six themes of Managing Search Operations (MSO)
      (1) Search is an Emergency: Get the teams out, treat the search seriously.
      (2) Search is the classic mystery-Continue investigations throughout the search.
      (3) Search for clues and the subject- use clue-aware teams, use trackers and dogs as clue finders.
      (4) Concentrate on aspects that are important and under your control. Plan your search around scenarios that you can affect.
      (5) Know if the subject leaves the search area-look at subject profiles, statistics, investigation, and, if appropriate, use containment.
      (6) Grid search as a last resort
   b. Remember your field experience. For each task sent out or decisions made, ask yourself if you were out in the field
      (1) Could I do what is asked?
      (2) Would I do what is asked?
      (3) Should I do what is asked?
   c. Remember your staff experience
8. Be nice, listen, and communicate

B. Expectations and the Customer
   1. Responsible Agent/Agency Administrator (RA/AA): Tact, ability to get along, fulfills RA objectives, enhances RA image, helps the sheriff get re-elected, finds subject.
   2. Family: Provides briefings, tact, honesty, finds subject.
   3. Staff: Provides clear objectives, able to teach roles, provide environment that

Union of Volunteer SAR Providers
Contract between field resources and management

* **Management will not let me stand around for more than 30 minutes.**
* **Management will give me enough tasks to allow me to become physically exhausted.**
* **Management will arrange for food and drinks so I don't really have to eat the 3-year-old sardines in my field pack.**
* **Management will provide me with a briefing or Subject Information Sheet.**
* **I agree to sign in and out of incidents.**
* **I will let management know when given an unreasonable task (if a dog handler I will complain loudly).**

1-6   Incident Commander for Ground Search and Rescue

protects staff, task idea guidance, answers questions.
4. Other SAR teams: Get teams out quickly, find subject, good briefing, good logistics.
5. Volunteers: Get teams out, feeling needed, briefings.
6. Subject: Found alive.

C. IC Oversight
1. While the IC may be in command, the successful IC always remembers that IC oversight (bosses) exist. In fact, the IC may report to several conflicting bosses at the same time.
2. Agency Administrator
   a. Establishes control factors
   b. Often has power to remove IC
   c. May have granted written delegation of authority
3. Local Government Officials
   a. Law Enforcement Agencies
   b. Board of Supervisors
   c. County Administrator
4. State Government Officials
   a. State IC certification
   b. Resource control
5. Federal Government Officials
   a. AFRCC
   b. Resource control
6. Group Oversight
   a. Removal as IC
   b. Removal from group

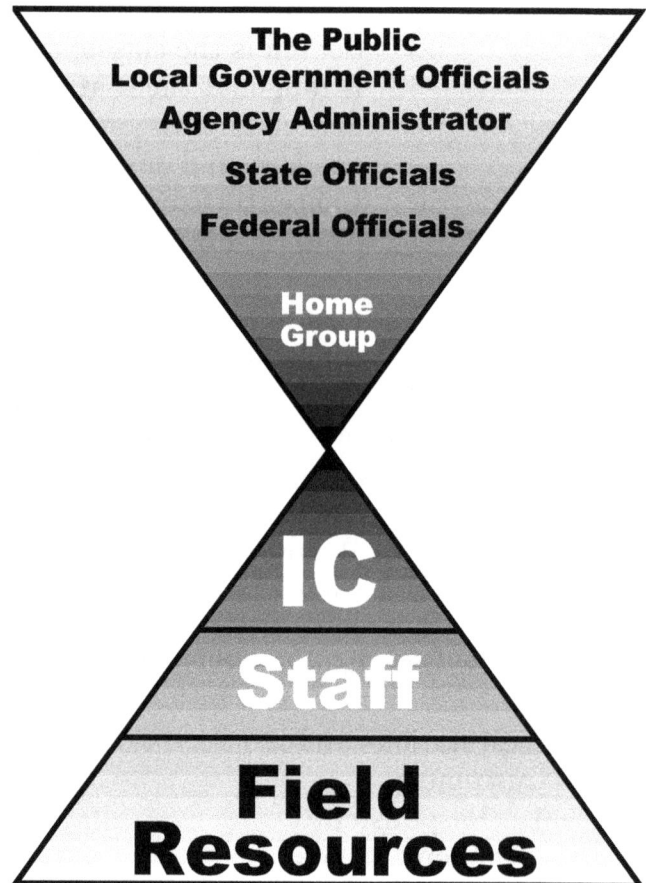

The Public
Local Government Officials
Agency Administrator
State Officials
Federal Officials

Home Group

IC

Staff

Field Resources

UNIT 2: INCIDENT COMMAND SYSTEM IMPLEMENTATION

I. **Review of ICS Parts**
   A. Ten common components that are the core of ICS

   B. Common terminology
      1. Use of plain English during radio traffic
      2. Common terms for Functions:
         a. Officers
         b. Chiefs
         c. Director
         d. Supervisor
         e. Leader
         f. Specific combination
      3. Common terms for resources
         a. Status
         b. Single resources, task forces, strike teams
      4. Common terms for facilities
         a. Command Post
         b. Incident Base
         c. Staging Areas, camps
         d. Helibases, Helispots
         e. Division
         f. Branch

   C. Functional Management/Modular Organization
      1. List Five Major Functions:
         a. Command
         b. Planning
         c. OPS
         d. Logistics
         e. Finance/Administration
      2. Top down responsibility
      3. Individual may be responsible for multiple functions

   D. Management by Objective
      1. ICS is management by objective. Therefore, no objectives = poor management.
      2. Objectives may be written or kept in the IC's head, but the IC cannot have an empty head.
      3. Objectives > Strategies > Tactics
      4. Common SAR Objectives:

Objectives:
☐ Demonstrate an understanding of the NIIMS Incident Command System as it applies to SAR and how the system can be adapted to any size incident.
☐ Demonstrate the ability to work within a unified command system with both SAR-trained and untrained incident commanders.
☐ Describe different structures for using the "PLOPS" function.
☐ Describe factors that influence the structure of ICS.

ICS is a tool - it helps the IC provide effective management but it does not guarantee success.

Objective	Strategy	Tactic
Keep the subject in the search area.	Establish containment.	Road patrols highway 11 and 610, Trail block AT
Search the high POA sectors to 80% PODcum.		
Locate the subject by the end of the operational period.		

E. Unified Command
  1. Define
  2. More detail later

F. Consolidated Action Plan
  1. Also known as Incident Action Plan (IAP), MSO text calls it a Search Action Plan (SAP).
  2. Can be written or unwritten (during the initial operational period)
     a. Written plan typically needed when:
        (1) Resources from multiple agencies are being used
        (2) Several jurisdictions are involved
        (3) Incident will require shifts of personnel and/or equipment
     b. Typical components of a written plan

G. Manageable Span of Control
  1. Rule-of-thumb range
  2. Accepted optimal
  3. Anticipate change
  4. Factors that alter

H. Modular Organization
  1. Define
  2. Section that changes the most from search to search

I. Integrated Communications
  1. Common communication plan everyone agrees to follow
  2. Plain text with no codes, minimal messages confined to essential traffic
  3. Role of pre-planning
  4. Includes radios, telephones, public address, mail bins, etc
  5. Will discuss more under making a communications plan

J. Designated Incident facilities
  1. Define
  2. Give example

K. Management of Tactical Resources
   1. Define
   2. Status definition
   3. Status tracking

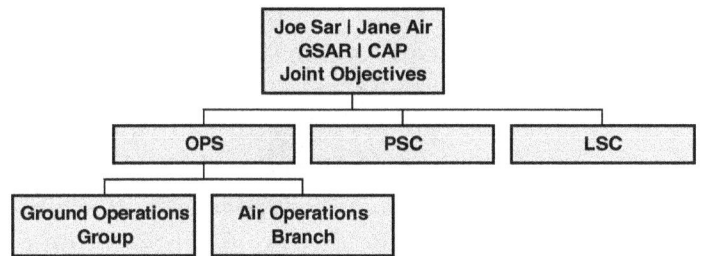

II. Unified Command

A. When required
   1. More than one department/agency shares management due to nature of incident
   2. Multi-jurisdictional
   3. Most common
      a. Virginia Civil Air Patrol Commander and a Ground State Incident Commander
      b. Public Service/Fire Captain and Ground State Incident Commander

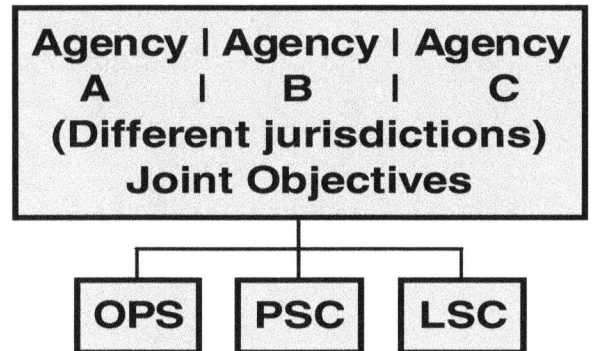

B. Single/Unified Command Differences
   1. Single command IC- solely responsible
      a. establishing objectives
      b. directly responsible for follow-through
      c. ensures all functional areas directed towards strategy
      d. OPS reports directly to IC
   2. In Unified Command
Individual members of unified command jointly decide
      a. objectives
      b. strategy
      c. priorities
      d. who will be OPS

C. Use of Unified Command with unprofessional "untrained" personnel
   1. SAR IC - Untrained IC
   2. Role of the OPS officer
   3. Use of dual structures
   4. Parallel structures
   5. RA/AA line to maintain control of state resources

D. Most Common use of Unified Command
   1. Selection of Air Incident Commander
   2. Selection of Ground Incident Commander
   3. Different structures for joint OPS Chief

Joe Sar | Jane Air
GSAR | CAP
Joint Objectives
— OPS — PSC — LSC
OPS → Ground Operations Group | Air Operations Branch

Agency | Agency | Agency
A | B | C
(Different jurisdictions)
Joint Objectives
OPS **PSC** **LSC**

Problem
RA/AA has given the Incident Commander responsibilities to the local fire chief before your arrival.
Solution?

III. Growth of a mission

A. Initial response: OHT
   1. IC, IS
      (1) IC politics, resource orders, planning, liaison, logistics, help develop tasks, send out tasks
      (2) IS- Operations (OPS), communications
      (3) Strong use of outside resources
       (a) Police- investigations/family
       (b) Rescue Squad-Medical Unit Leader
       (c)   Rescue/Fire- radio operator
       (d) Local-Logistics (food, shelter, water)

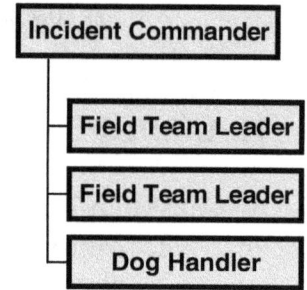

```
┌──────────────────────┐
│  Incident Commander  │
└──────────┬───────────┘
           ├──┌────────────────────┐
           │  │  Field Team Leader │
           │  └────────────────────┘
           ├──┌────────────────────┐
           │  │  Field Team Leader │
           │  └────────────────────┘
           └──┌────────────────────┐
              │    Dog Handler     │
              └────────────────────┘
```

   2. IC, 2 IS
      a. IC-IC stuff
      b. IS- OPS
      c. IS- Investigation /family-plans
      d. Other structures

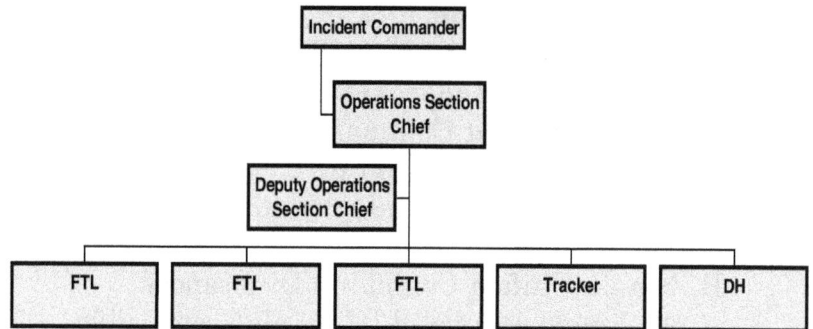

```
                    ┌──────────────────────┐
                    │  Incident Commander  │
                    └──────────┬───────────┘
                    ┌──────────┴───────────┐
                    │  Operations Section  │
                    │        Chief         │
                    └──────────┬───────────┘
         ┌────────────────────┐│
         │  Deputy Operations ├┤
         │    Section Chief   │
         └─────────┬──────────┘
   ┌──────┬──────┬─┴────┬──────────┬──────┐
┌──┴──┐┌──┴──┐┌──┴──┐┌───┴────┐┌────┴───┐
│ FTL ││ FTL ││ FTL ││Tracker ││   DH   │
└─────┘└─────┘└─────┘└────────┘└────────┘
```

   3. PLOPS (Plans and Operations) Function: Different Structures
      a. "No Plops"-
      b. True Plops- 2 staff
      c. Modified Plops

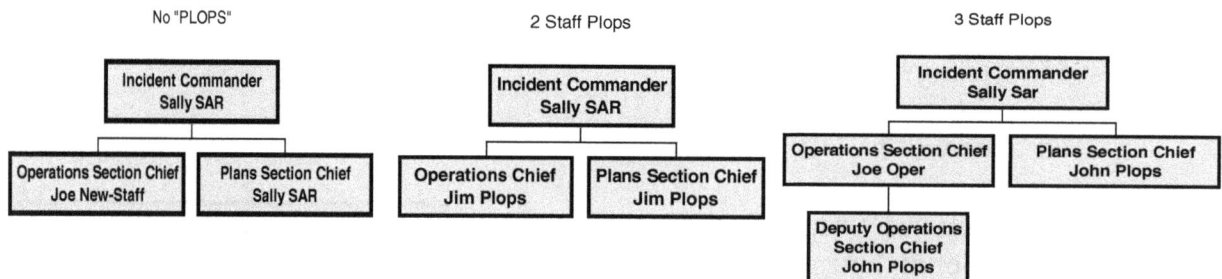

No "PLOPS"

```
        ┌────────────────────┐
        │ Incident Commander │
        │     Sally SAR      │
        └─────────┬──────────┘
    ┌─────────────┴─────────────┐
┌───┴─────────────┐ ┌───────────┴──────┐
│Operations Section│ │Plans Section Chief│
│  Chief           │ │   Sally SAR       │
│  Joe New-Staff   │ │                   │
└──────────────────┘ └───────────────────┘
```

2 Staff Plops

```
        ┌────────────────────┐
        │ Incident Commander │
        │     Sally SAR      │
        └─────────┬──────────┘
    ┌─────────────┴─────────────┐
┌───┴──────────┐ ┌──────────────┴───┐
│Operations    │ │Plans Section Chief│
│  Chief       │ │   Jim Plops       │
│  Jim Plops   │ │                   │
└──────────────┘ └───────────────────┘
```

3 Staff Plops

```
            ┌────────────────────┐
            │ Incident Commander │
            │     Sally Sar      │
            └─────────┬──────────┘
      ┌───────────────┴───────────────┐
┌─────┴──────────────┐    ┌────────────┴──────┐
│Operations Section  │    │Plans Section Chief │
│  Chief             │    │   John Plops       │
│  Joe Oper          │    └────────────────────┘
└─────────┬──────────┘
    ┌─────┴──────────────┐
    │ Deputy Operations  │
    │   Section Chief    │
    │   John Plops       │
    └────────────────────┘
```

B. Initial Response/limited Growth Quick Response Team (First shift)

   1. IC- IC stuff.
   2. IS- OPS.
   3. IS/mature person- Investigations.
   4. IS- Plans (medical plan, evacuation, communications, Matteson, sectors, objectives, next shift resources, logistics, tactical planning, deputy OPS).
   5. Qualified- Base Radio Operator, Communication Unit Leader.
   6. Sign-in/Staging.

C. Critical Functions

Function	Often performed by
Command	IC
Operations	OPS, OG, PLOPS
Plans	IC, PSC, OG, PLOPS
Logistics	LSC, IC, PSC
Communications	OPS, Communications Unit Leader
Clues	OPS, PSC, CUL, SITSTAT, Clue Director
Investigations	Investigator, Law Enforcement, PSC
PIO	IC, AA/RA, PIO

IV.   **Different Structures**

A. Size Considerations
   1. Reasons to increase staff size
      a. Better tracking for safety
      b. Get the teams out faster
      c. Better documentation for long-term searches
      d. Head off potential problems
   2. Reasons to decrease staff size
      a. Need to send resources into the field
      b. Physical facilities cannot maintain a large staff
      c. Leaner staffs are often more efficient
   3. Options
      a. Expand and contract staff size as needed
      b. Allow staff to rest/nap when possible
   4. New problems
      a. Will always be criticized
      b. Unqualified, incompetent who want to be staff
      c. Members worried about recertification requirements
      d. Members who just want to stay in base

> **Problem**
> Several Dog teams are waiting to be debriefed. Your briefing person is already swamped trying to get additional teams back out into the field.
> **Solutions?**

B. Structural considerations
   1. Differences mostly in operations
   2. Factors that influence
      a. Who is available
      b. Number of searchers
      c. Size of search (geographic)
      d. Communications
      e. Terrain
      f. Transportation
      g. Logistics
      h. Philosophy of IC
   3. Rules of Thumb:
      a. Bob's 1:10 ratio
      b. Greg's 1:5 ratio
      c. Staff will always look busy
      d. Day vs Night for busy work
   4. Branches
      a. Used two different ways in ICS. Can be based upon function or a grouping of divisions
      b. Common examples of functional branches: Dog, Ground, Horse, Air.
      c. Branch supervisor located at base CP
      d. Different frequencies often assigned
      e. Use of branches for organizing divisions only occurs on the largest of

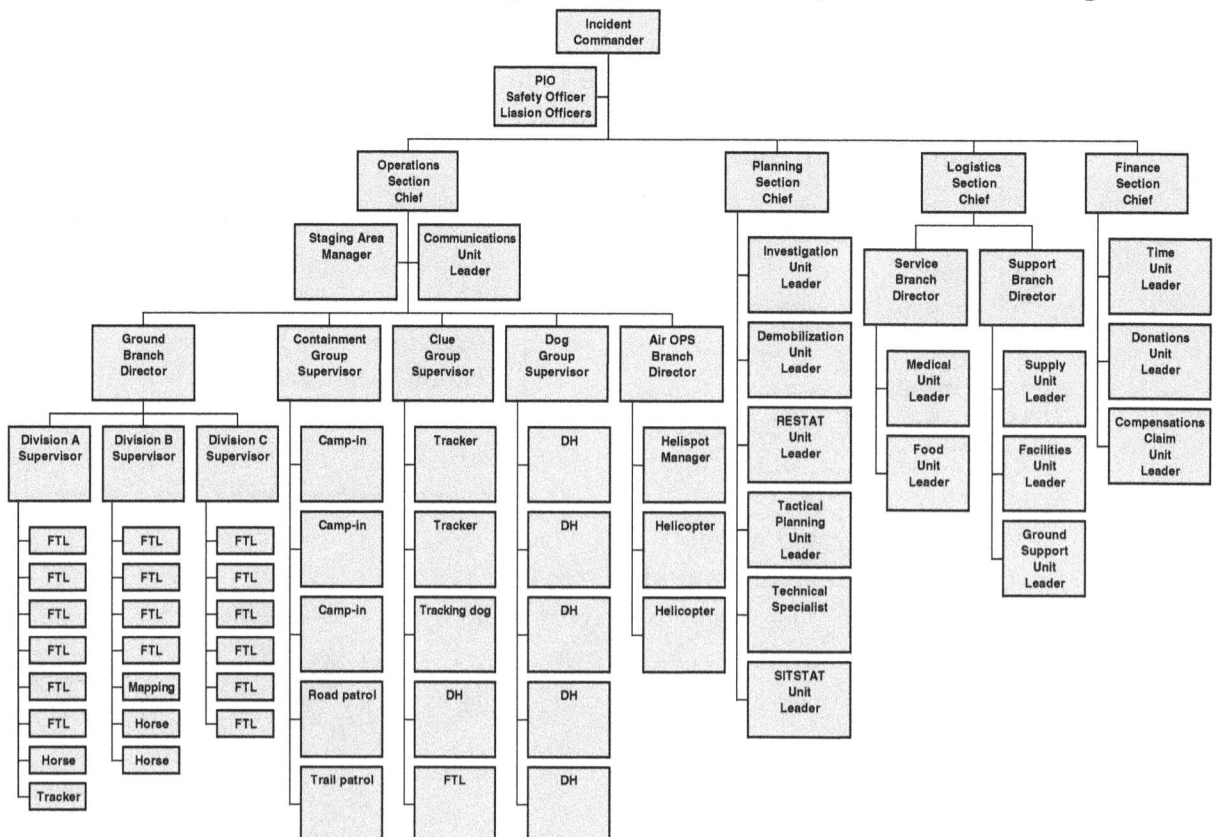

No matter how little useful work is being accomplished, the staff will always look busy.

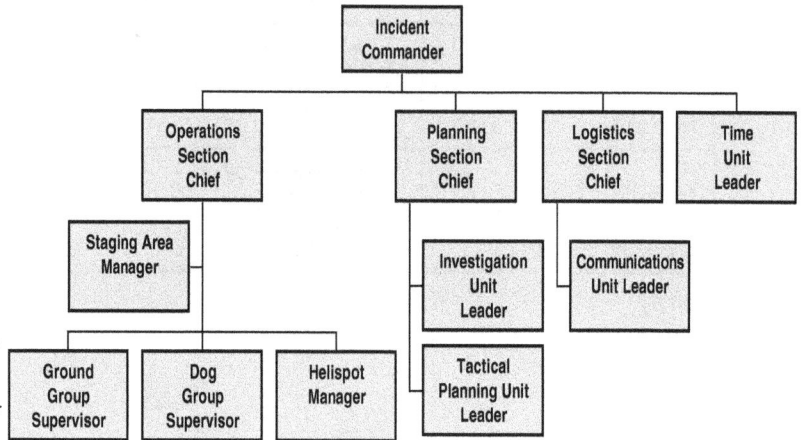

searches. Uncommon for search and rescue incidents.

5. Divisions
   a. Based upon geography
   b. Division supervisor usually located at remote division
   c. Different frequencies for each division
   d. Command frequency often used
   e. Several other considerations

6. Examples of different structures:

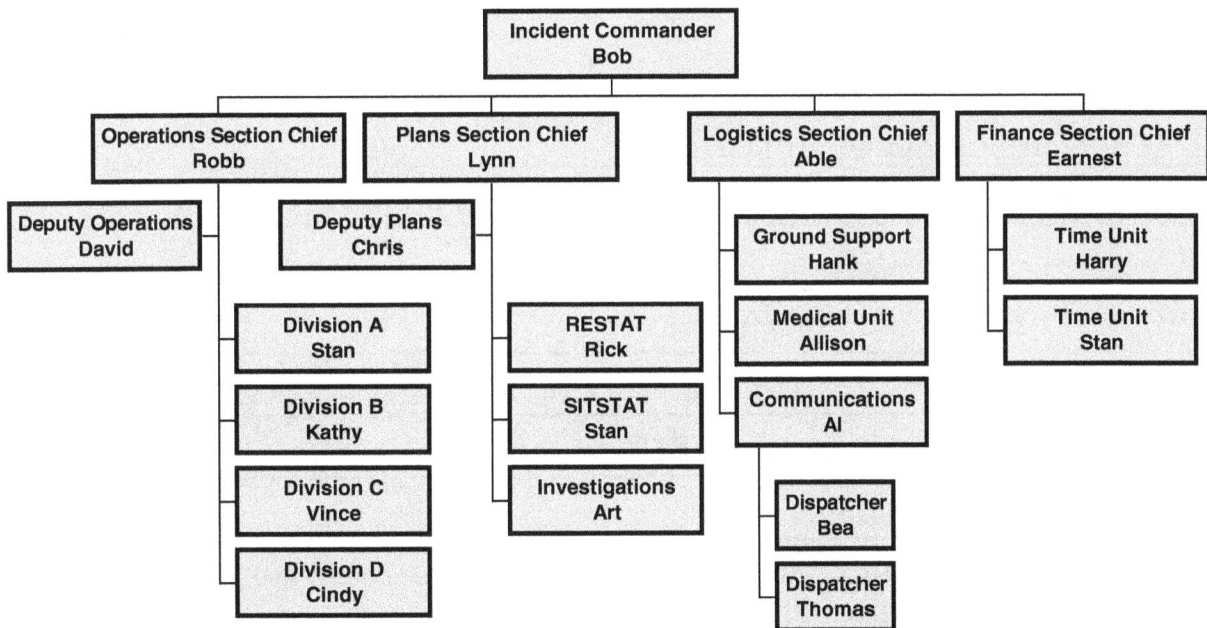

```
                          Incident Commander
                                Bob
        ┌─────────────┬──────────────┬──────────────┐
  Operations Section    Plans Section   Logistics Section   Finance Section
     Chief Robb          Chief Lynn        Chief Able         Chief Earnest

  Deputy Operations    Deputy Plans     Ground Support       Time Unit
      David               Chris             Hank               Harry

     Division A          RESTAT         Medical Unit         Time Unit
       Stan               Rick            Allison              Stan

     Division B          SITSTAT        Communications
       Kathy              Stan               Al

     Division C       Investigations      Dispatcher
       Vince              Art               Bea

     Division D                          Dispatcher
       Cindy                              Thomas
```

Geographic Divisions

```
                          ┌─────────────────────────┐
                          │   Incident Commander     │
                          │          Bob             │
                          └─────────────────────────┘
        ┌──────────────────────────┬──────────────────────────────────┐
┌─────────────────────┐   ┌──────────────────────┐   ┌─────────────────────────┐
│ Operations Section  │   │  Plans Section Chief │   │ Logistics Section Chief │
│   Chief  Rita       │   │         Rita         │   │    John Fire Chief      │
└─────────────────────┘   └──────────────────────┘   └─────────────────────────┘
  ┌──────────────────┐        ┌──────────────────┐       ┌─────────────────────────┐
  │ Staging Area     │        │  Investigation   │       │   Medical Unit Leader   │
  │ Manager  Karen   │        │       Wes        │       │     BridgewaterRS       │
  └──────────────────┘        └──────────────────┘       └─────────────────────────┘
  ┌──────────────────┐                                   ┌─────────────────────────┐
  │ Deputy Operations│                                   │   Food Unit Leader      │
  │     Theresa      │                                   │     RS Auxillary        │
  └──────────────────┘                                   └─────────────────────────┘
┌──────────────┐ ┌──────────────┐ ┌──────────────┐       ┌─────────────────────────┐
│ Ground Branch│ │  Dog Branch  │ │ Diver Branch │       │  Facilities Unit Leader │
│    Rita      │ │    Steve     │ │     Jan      │       │      Fire Dept.         │
└──────────────┘ └──────────────┘ └──────────────┘       └─────────────────────────┘
                                                         ┌─────────────────────────┐
                                                         │  Ground Support Leader  │
                                                         │     Cave Springs RS     │
                                                         └─────────────────────────┘
                                                         ┌─────────────────────────┐
                                                         │     Communications      │
                                                         │        Theresa          │
                                                         └─────────────────────────┘
                                                         ┌─────────────────────────┐
                                                         │  Base Radio Operator    │
                                                         │        ASRC FTM         │
                                                         └─────────────────────────┘
```

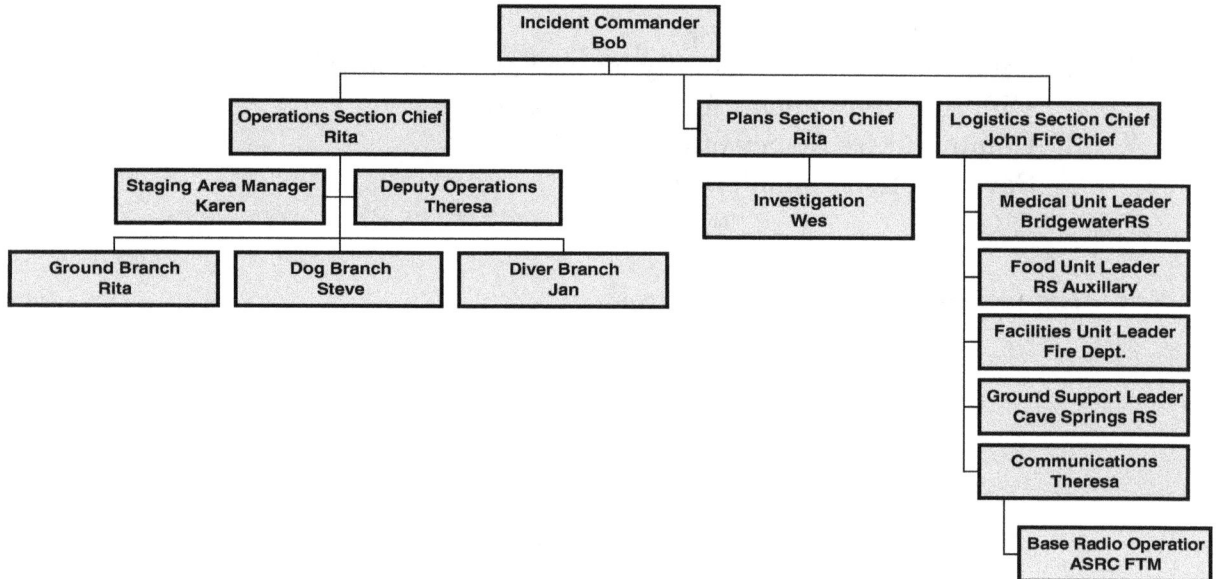

Functional Branches

V. ICS Positions and Qualifications

ICS Position	Position Training Qualification or Equivalent
IC	IC
PIO	IC/IS/PIO training
Safety	IC/IS/Rescue Specialist (RS)
Liaison	IS/experienced FTL, sensitivity
PSC, OPS	IC/IS
LSC	LSC, if only food, facilities, and transport; use local resources with authority
FSC	Accounting and cost/benefit analysis, familiar with any required paperwork
Div. Sup	IC/IS
Branch	IS/ Experienced FTL/Resource Specialist
Air OPS	CAP/NIMMS Qualified
CUL	IS/CUL/CAP qualified
Investigation	IS/Maturity
Briefer/Debrief	FTL

UNIT **3**: OPERATIONAL TACTICS

I. General Task Considerations

A. Type of resources immediately available
B. Recall field experience in making decisions
C. Highest POD in highest POA
D. Reflex tasks
E. Resource mixing
F. Waiting for resources
   1. PLS resources
   2. Man-tracking, Tracking dogs
   3. Need for action
G. Need to specify detailed instructions on Task Assignment Form (TAF)
H. Task Length
   1. Desired length
   2. Assumptions: Transport to search area, need to stay in radio contact
   3. Problems if task too long or short

Objectives:
Demonstrate an understanding of SAR resources including appropriate and inappropriate uses of the following types: ground teams, air-scent dogs, trailing/tracking dogs, mantrackers, mounted horses, helicopters, fixed wings, ATVs, bicycles, and kayakers.

II. Ground Teams

SEARCH TACTICS

A. Linear, Scratch

Objective: Cover high probability areas, quickly, using minimal resources

1. 2-4 persons on team
2. Follows travel aids
3. FTL field promotions based upon several factors and require IC permission
4. Navigation greatest challenge
5. Clue awareness important
6. Start task at highest POA if possible
7. Downhill if possible
8. Consider transportation for start/end point
9. Clear start point aids navigation
10. Concentration time factor
11. Concentration better at start
12. Remind shouting, signaling
13. Specify if spread desired
14. Easily combined with sound and tracking techniques

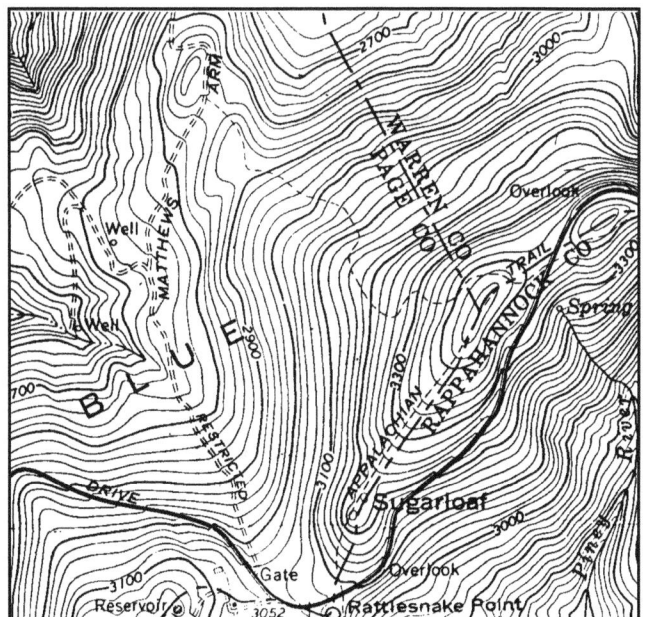

Fig 3.1 Linear Task

B. First check, Hasty
Objective: Check the immediate area, building, and specific areas of high probability of area.

1. Typically conducted in immediate area of Point Last Seen (PLS) or Last Known Position (LKP)
2. Critical not to destroy PLS and look for direction of travel and tracks
3. Typically conducted by first on-scene resource
4. If PLS around building make a thorough check of all nearby buildings
5. Should repeat on a periodic basis

C. Wandering
Objective: Cover large amounts of terrain with minimal resources for responsive subject.

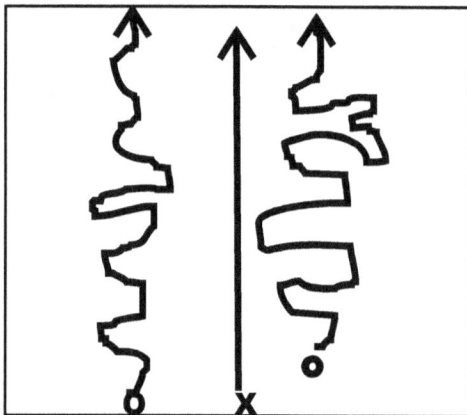

Fig 3.2 *Anchor Wandering*

1. 2-3 person team assigned to sector
2. Research shows efficient as Grid
   a. Family finds
   b. Dog evidence
3. Excellent navigation needed
4. Field Team Leader (FTL) intensive
5. Good for high probability areas if run out of dogs
6. Limited use at night
7. Easily combined with "sound task"
8. Two techniques: Anchor (**Fig 3.2**)& Clue Aware Team (CAT) (**Fig 3.3**)

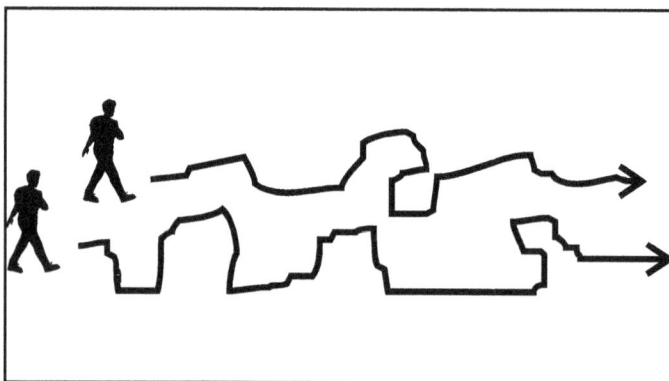

Fig 3.3 *"CAT" Wandering*

D. Modified Trail/Road Sector (**Fig 3.4**)
Objective: Quickly cover the area of highest probability with minimal resources.

1. 3-10 person team
2. Best for subjects found short distances off roads and trails
3. Navigation is not as critical
4. Specify distance from road for desired coverage
5. Techniques

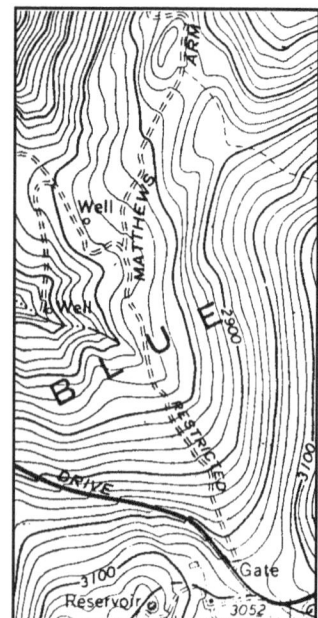

E. Sweep (**Fig 3.5**)

Objective: Non-thorough technique to efficiently cover search areas.

1. 6-12 people (>10 problems start)
2. Size of team factors
   a. number of FTLs
   b. number of volunteers
   c. often used to "use up" volunteers
   d. terrain, type of search
   e. responsiveness of subject
3. Consider transportation
4. Good boundaries important, especially at night
5. Artificial boundaries of flagging tape
6. Speed desired
7. Sweep vs Grid vs Briars
8. Often used early in search in high POA sectors
9. Attention span of volunteers highly variable
10. Specify special equipment needs (flagging tape)
11. Specify range of spacing
12. Discuss critical spacing concept
13. Give targeted POD, expect less, don't expect more
14. Perception problem when mapping task
15. Techniques
    a. contour
    b. drainage
    c. bearing, sectors

F. Grid (**Fig 3.5**)

Objective: Thorough technique to raise POD and look for unresponsive subjects.

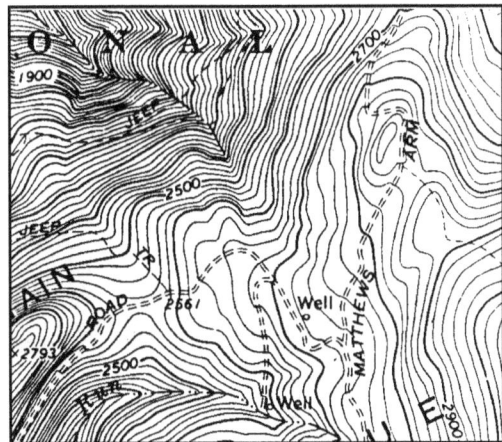

Fig 3.5 *Sweeps and Grids*

1. 8-16 people (10 still ideal)
2. Tighter spacing
3. Specify spacing
4. Specify critical spacing
5. Specify targeted POD
6. Solid boundary required
7. Alternative boundary
8. Staggered starts
9. Unclear the difference with sweeps sometimes in dense areas
10. Competent flankers required
11. It becomes more important to use an experienced FTL with more untrained volunteers

G. Expanding Circle
Objective: Thorough check around PLS, LKP, or clue to locate clues, direction of travel, or subject.

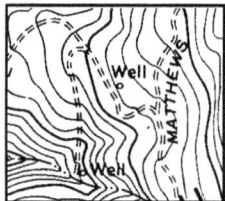

Fig 3.6 *Expanding Circle Task*

1. 2-10 person team
2. Effective only for small area
3. Typical follow up to clues or PLS
4. Experienced team required to avoid destroying clues/tracks
5. Specify maximum expansion
6. May be important to protect PLS with barrier tape
7. Ideally a mantracking task or with a team that has members with tracking skills

Fig 3.7

H. Mapping Task
Objective: Scouting task to update or verify maps.

1. 2-3 person team
2. FTL with excellent navigation skills
3. Special equipment: Global Positioning System, Compass, Altimeter, several copies of map, topo under acetate
4. Scouting from a helicopter or fixed wing useful for mapping flagging tasks

I. Flagging Task (**Fig 3.8**)

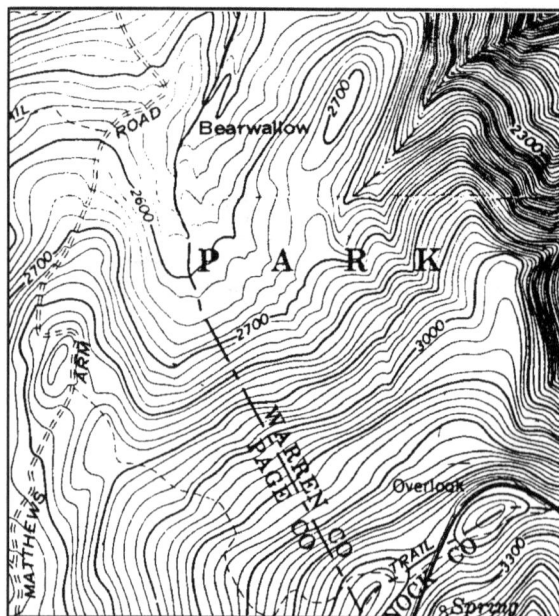

Fig 3.8 *Flagging Task*

Objective: Lay down a trail of flagging tape.

1. 2-3 person team
2. FTL with excellent navigation skills
3. Ideally should be performed during day but often done at night, consider day break head start
4. Make sure flagging lines labeled and color coded
5. Task timing becomes critical
6. Used for marking route of evacuation
    a. Dense flagging with long tails
    b. If at night/fog consider solid streamer
    c. Navigation critical, consider advance scouting

J. Containment

Objective: Keep subject within search area. Locate subject on roads or trails.

    1.   Decide early if required

    2.   Decide value: Analyze type of search and terrain

    3.   Types

Fig 3.9

        a.   Road Patrol
          (1) Law enforcement
          (2) Rescue squad
          (3) Volunteers
          (4) Use of sirens, lights, PA system

Fig 3.10

        b.   Trails
          (1) Horses
          (2) Hasty
          (3) Real FTL
          (4) Trackers/Tracking dogs
          (5) ATVs/Bikes

Fig 3.11

        c.   Camp-ins
          (1) Pairs
          (2) Scheduled relief required
          (3) Fires good for attraction and morale
          (4) Ensure properly equipped
          (5) Trailheads, trail junctions
          (6) Surveillance points

        d.   Surveillance
          (1) Some success stories
          (2) Level headed
          (3) Well equipped
          (4) Solid FTL
          (5) Warn against direct action
          (6) Good communication
          (7) Special equipment, sources

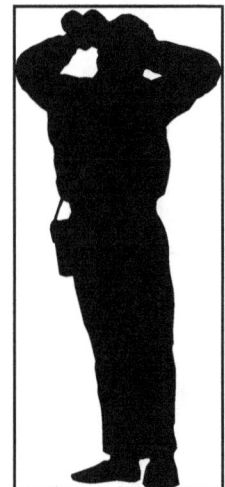

Fig 3.12

        e.   River Containment
          (1) Used for suspected drowning victims in smaller shallow rivers or streams
          (2) Dive team installs "construction" fence across stream
          (3) Fence checked several times a days for victim
          (4) Make sure fence far enough downstream
          (5) Make sure river not at flood stage or fence usually destroyed by floating debris
          (6) Take down fence at some point

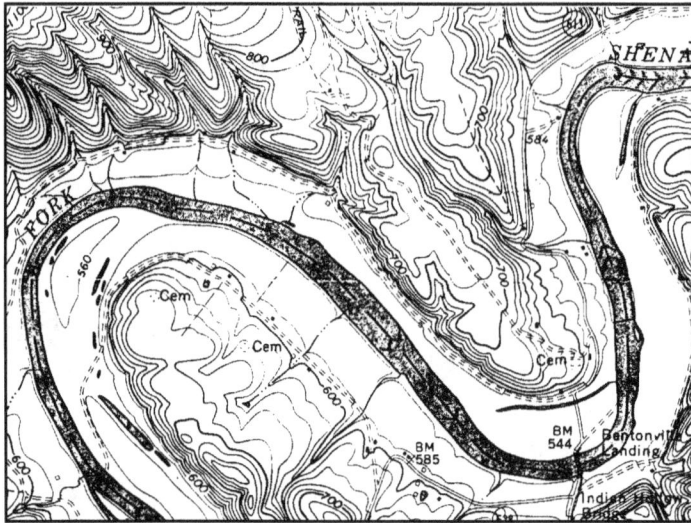

Fig 3.13 *River Containment*

K. River Surveillance
Objective: Locate subject or clues of drowning incidents.

    1. Linear task along river or stream bank to see if body has surfaced or can be seen. Incidents often occur when streams or rivers are at or near flood stage.

    2. Hazard Assesment and mitigation

      a. Hazards include: falling into water, slippery banks, undercut river banks, debris fields, displaced snakes. Changes in water course will not appear on topo map.

      b. Consider issuing PFD and throw bags to teams

      c. Make sure team does not have to cross major streams entering river

      d. Make sure all team members can swim

      e. Consider aborting tasks at night if looking for deceased subject and hazards are high

Fig 3.14

    3. Task Considerations

      a. Issue binoculars to teams

      b. Team size can be small. Task will be slow and rough due to debris.

      c. Bank search also involves some area search as water levels lower

      d. Remind members to look in trees up to water level for deceased subjects and above water line for survivors

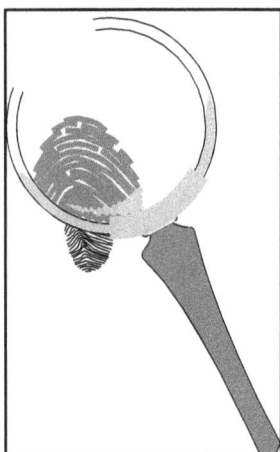

Fig 3.15

L. Investigative
Objective: Collect and disseminate information to assist in the search investigation.

    1. Can be combined with most other search tactics.

    2. Posting fliers and citizen interface

    3. More general area maps generally required

    4. Review investigative and interviewing techniques with FTL if required

    5. Ensure team presents a professional appearance

    6. Ensure excellent documentation

    7. Clearly state expected thoroughness of task

    8. Consider Staff overview flights with helicopters

    9. Consider using family members to post fliers

M. Communications
Objective: Provide manual communication relay if repeater not available.
1. Function as manual relay to overcome radio-dead zones
2. Follows suggestions of camp-ins
3. Ensure personnel selected have excellent communications skills
4. Attempt to stage from area that can be reached by vehicle
5. Issue extra batteries, ensure vehicle has full tank and a good battery

N. Rescue-Related Tasks
Objective: Safely remove subject from the field.
1. Medical Team
    a. Consists of medic, equipment, and assistants. Should be predetermined in medical plan.
    b. Dispatched once location and best route selected. Often moves in front of incoming evacuation team.
    c. Assess, stabilizes, and prepares patient for evacuation

Fig 3.16

2. Flagging Team
    a. Consists of flagger and assistant
    b. Described under flagging tasks previously
    c. Team often made from part of find team or other incoming field teams
    d. Lots of flagging tape required
    e. Avoid sending a single person if hazardous terrain

3. Route Preparation
    a. Typically follows flagging team
    b. Team usually made from other field resources
    c. Help clear out vegetation and prepare route

4. Rescue Rigging
    a. Technical team to prerig technical ropes in front of evac team
    b. Selection of trained technical leader critical
    c. Consider a separate safety officer to check work
    d. Team may require several assistants to bring in equipment

5. Evacuation Team
    a. Make sure team is equipped for extended time in outdoors
    b. Make sure sufficient personnel exist
       (1) check personnel equipment, physical conditioning
       (2) ability to distribute team equipment
    c. Appoint strong leader who is well qualified in both technical aspects and leadership
    d. Carefully select members and avoid general rush to field
    e. Carefully document team members
    f. Consider several waves of team members

III. Air-Scent Dogs

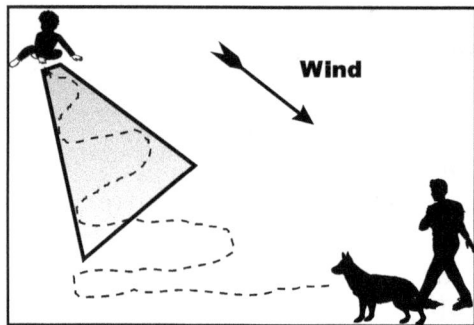

Fig 3.17 *An Air-scent dog working the scent cone.*

A. General Considerations
1. How an air-scent dog works:
   a. What it smells (**Fig 3.17**)
   b. Dog olfactory physiology
   c. What an alert means
   d. How a dog follows up an alert
2. 50% of dog finds during sector/grid searches
3. Able to use most dog handlers in staff positions

4. Air-Scent Search Dog specializations
   a. Wilderness
   b. Water
   c. Disaster
   d. Cadaver/evidence
   e. Scent discriminating
   f. Need to ask handler
   g. If working with an unknown team need to ask how they train and what standards they use
5. Resource used for rapid field deployment, often self deploy
6. Prebriefing a team technique
7. Environmental conditions that favor the dog's ability
   a. Night
   b. Light rain or drizzle, fog
   c. Light winds from a consistent direction
   d. Temperatures of 45-75° F
   e. Open woods or fields
   f. Others
8. Environmental conditions that **do not** favor the dog:
   a. Low/high humidity
   b. High temperatures
   c. Low temperatures (below 32°F) if the subject is dead
   d. No wind
   e. Constantly changing light winds
   f. Thick foliage
   g. Highly mountainous terrain with large boulders and squirrely wind
   h. Obstacles that impede dogs (**Fig 3.18**)

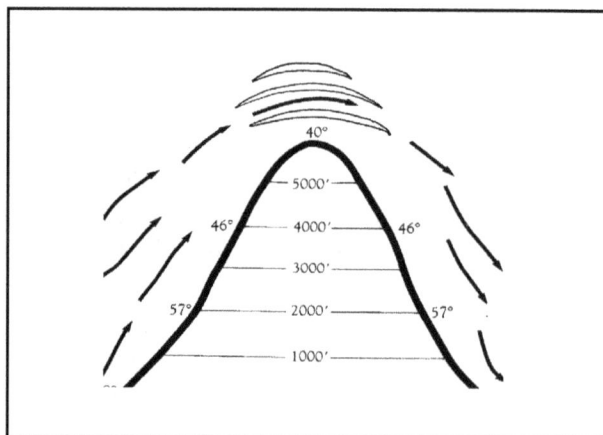

Fig 3.18 *Air circulation around an obstruction.*

B. Wind issues
  1. Air clearance time
     a. 30 minutes- 2 hours
     b. Dog handlers like to know who is in adjoining sectors
     c. Don't rule out subject too quickly when teams find each other
     d. Physical objects and scent pooling

  2. Microcirculation and mountain meteorology facts.
     a. If large front moving through predominant wind direction established by the front
     b. Need to understand both wind and temperature circulations
     c. Examples
     d. Factors that support micro-climates
        (1) Clear skies, light winds
        (2) Period between storms
        (3) Weak high-pressure systems
     e. Atmosphere contains both horizontal and vertical movements
        (1) First 3000 feet of elevation contains the most unstable air
        (2) Winds near the surface increase during the morning. As air rises, it slows down the speed of upper winds.
        (3) Think of the vertical air movements a candle produces
        (4) As the day progresses wind speeds at the surface increase while the speed on the mountaintop slows down
        (5) At night the ground cools, air becomes stable, vertical convections stop, surface winds slow down, mountaintop winds increase
        (6) Convection columns are similar to bubbles, resulting in gusty winds to calm winds. Difficult to predict gusts.
     f. Local Winds
        (1) Cause of local winds is differential heating and cooling of topology.

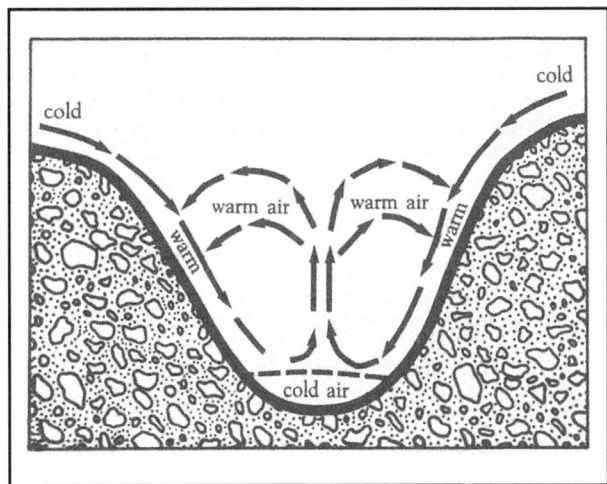

Fig 3.19 *Air circulation in a valley on a clear night*

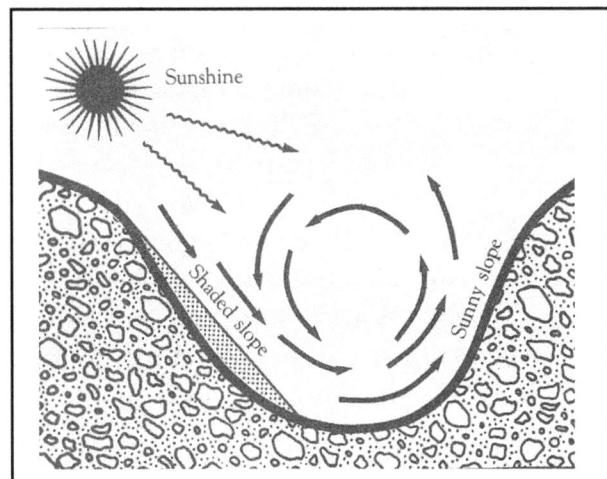

Fig 3.20 *Air circulation in a valley on a clear day when one slope is shaded*

(2) Slope Winds
  (a) **Winds blow upslope during the daytime**
    i)   Heated air during the daytime does not hug the ground but peels off and forms convection columns, usually following along a heated vertical object such as a tree or building.
    ii) Slope which receives the most sunlight will have the strongest winds

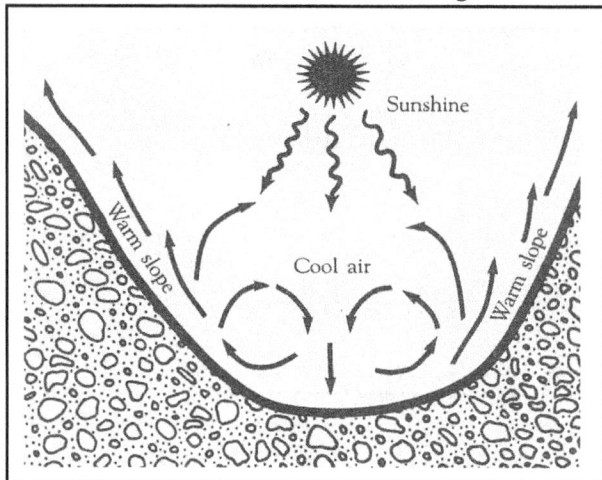

Fig 3.21 *Air circulation in a valley on clear days*

iii)**Winds blow downslope during the nighttime**
a)True on calm nights when the air is clear and dry
b)Winds are light, a few miles an hour, and steady on moderate slopes
c)On steeper slopes the air speed increases
d)Steep slopes also experience "air avalanches"
e)Sharp valleys have less pooling and winds since the sides reradiate to each other and decrease temperature differentials
f)Thermal belts can have temperatures that are 20°F higher

  g) Low clouds obliterate local winds

C. The Dog Handler (DH) as an FTL
  1. Most teams train to the FTL level except for semi-technical skills and leading sweep/grid tasks
  2. Most carry personal or group radio
  3. Most excellent at navigation
  4. Most excellent at family and press relations
  5. Most provide valuable input for staff decisions
  6. Most prefer trained walker at the Field Team Member (FTM) level
  7. Walker considerations
    a. Save FTL for other tasks if possible. However, keep in mind weaker navigation skills of some handlers.
    b. Try to use FTMs with strong skills
    c. Walker needs good radio and navigational skills
    d. Good physical shape important
    e. Walker may ask questions but needs good judgement to avoid chatter
    f. Advantages and disadvantages of a local resources/law enforcement as walkers

SEARCH TACTICS

D. Sector search (Grid Search)(**Fig 3.22**)
Objective: Thorough technique (wind permitting) to cover large area with minimal resources.

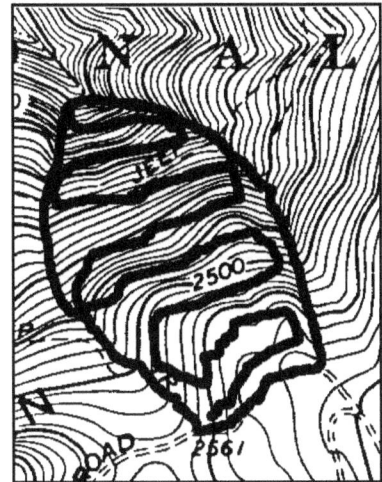

Fig 3.22 *Air-Scent Grid pattern*

1. Assign good boundaries to sector
2. Rule of thumb for boundary size
3. Adjust size for night
    a. Smaller
    b. More distinct boundaries
4. Dog will work segment based upon wind direction
    a. Often work perimeter first
    b. Second, grid pattern established
5. Cover large areas with minimal resources
    a. Evasive subjects
    b. Subjects in thick foliage
    c. Subjects found by placement of any resources into sector
6. Debriefing needs to plot exact route of task

E. Modified Trail/Sector
Objective: Quickly cover the area of highest probability with minimal resources.

1. Trail forms primary sector boundaries (see **Fig 3.4**)
2. Secondary boundary dependent upon subject profile, resources, and topography
3. Search pattern dependent upon wind direction
4. Used to look for subjects that do not travel great distances off trails and roads

F. Expanding Circle (**Fig 3.23**)
Objective: Thorough check around PLS, LKP, or clue to locate clues, direction of travel, or subject.

Fig 3.23 *Expanding Circle*

1. Appropriate locations
    a. PLS, LKP
    b. Physical clue
    c. High-likelihood area
2. Useful with subjects that do not travel far from PLS
    a. Alzheimer's patients
    b. Despondents
    c. Murders
    d. Some lost children
3. Losses effectiveness the larger the circle. Must be track aware.
4. Used as an initial task or a follow-up to a clue

Fig 3.24 *Technique A*

Fig 3.25 *Technique B*

G. Coordinated Dog/Ground
Objective: Provide a high POD and mixture of resource types in a single pass in a time critical, high-POA sector.
1. Covers a large area in one pass
2. Different types of resources complement each other
3. Requires increased cooperation and training
4. Technique A - Involves a single dog team working with a sweep team(**Fig 3.24**).
5. Technique B - Involves multiple dogs with a sweep team(**Fig 3.25**). Can cover tremendous amount of territory quickly, but requires practice before successful implementation.

H. Coordinated Dog/Horse
Objective: Cover areas or trails using both horses and air-scent dogs.
1. Handler usually on horseback
2. Dog allowed to range greater distances
3. Terrain must allow horseback riding
4. Allows coverage of greater areas
5. Allows greater viability for handler and combines advantages of both techniques

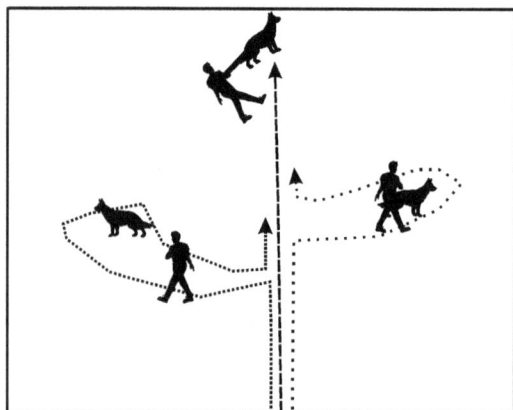
Fig 3.26 *Coordinated Dog Task*

I. Coordinated Dog/Tracking Dog
Objective: Combine the advantages of air-scent dogs and tracking dogs into one task.
1. Typically used to cover a trail
2. Start tracking dog first (**Fig 3.26**)
3. After a 15-30 minute head start use two staggered flanking air-scent dogs
4. Air-scent dogs often able to pick up subject if tracking dogs misses

J. Coordinated Dog/Mantracker
Objective: Combine the advantages of air-scent dogs and mantrackers into one task.
1. Air-scent dog team's walker is a trained tracker
2. May be used with a wide variety of dog-task types
3. Tracker able to pick up prints on trails, at track traps, and trail crossings
4. Able to deploy air-scent dog and tracker into areas of high probability simultaneously
5. Tracker fulfilling more of a reconnaissance role

K. Scratch and Sniff (**Fig 3.27**)
Objective: Scout, locate, and search non-populated areas in an urban area not indicated on topographic maps.
1. Used in urban searching
2. Best if dogs have had disaster training
3. Problem of lots of scent
4. Problem locating woods on city topos
5. Technique
    a. DH assigned large sector
    b. Drives around and locates patches of woods
    c. DH searches woods
    d. Careful map plotting required
6. Urban search considerations

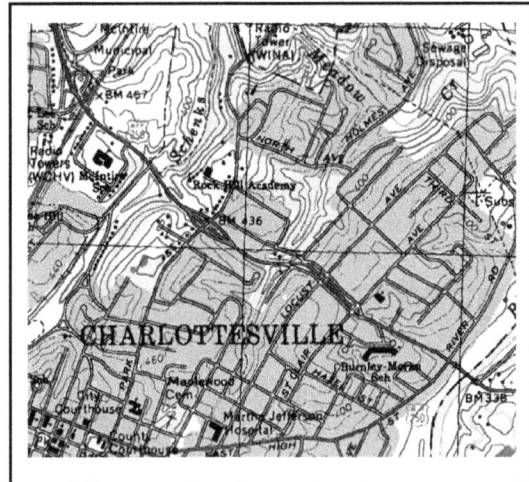

Fig 3.27 *Scratch and Sniff in an Urban area.*

L. Water searching
Objective: Locate scent of subject underwater in order to direct divers.
1. Brief history
2. Scent description
3. Differences in alerts
4. Safety concerns more stringent since looking for a status three. Carefully evaluate need for night tasks
5. Need good boat, stable platform, fishing/work boat
6. Prefer professional boat operator (Inland Game and Fisheries)
    a. Has law-enforcement powers on water
    b. Knows local waters
    c. Highly trained and safe boat operator
    d. **MUST have an electric motor or oars**
    e. Some dog handlers may use canoes
        (1) Special safety concerns
        (2) Knowledge of local water critical
7. Ensure use of PFD's
8. Shoreline operations possible
9. Shoreline support may include swiftwater rescue teams, spotters, spotters with throw bags

IV. Trailing and Tracking dogs

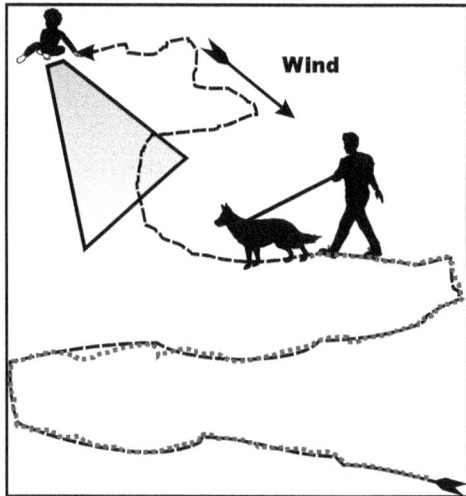

Fig 3.28 *Tracking dog follows subject's footsteps by smelling crushed vegetation*

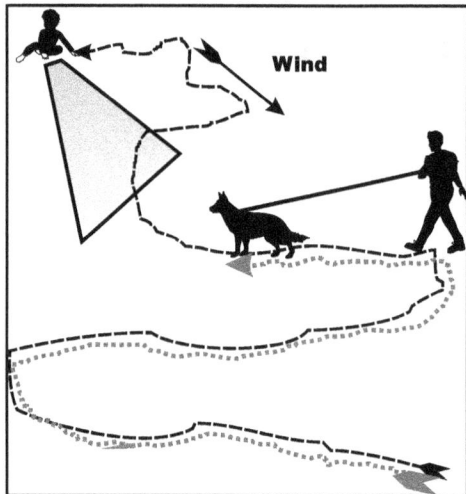

Fig 3.29 *Trailing dog follows subject's scent as it falls influenced by local wind*

A. Tracking versus trailing dogs
1. Most common type
2. Tracking: Dog follows crushed vegetative scent caused by footsteps (**Fig 3.28**)
3. Trailing dog (blood hound): Follows human (scent) cells that have fallen to the ground (may be carried by wind a short distance) (**Fig 3.29**)
4. Advantages of tracking:
    a. Often available at the county level through corrections department
    b. Able to locate subject quickly if track is successful
    c. Able to provide a direction of travel
    d. Can work at night and with other resources
    e. Does **NOT** require a scent article
5. Advantages of trailing
    a. Same as tracking dog except a scent article is required
    b. Allows more contamination of PLS and search area
    c. Scent discrimination allows trail to be started anywhere the subject traveled
    d. Scent discrimination allows verification of located clues
6. Disadvantages of tracking
    a. Dogs must be used before anyone crosses path since they will track the last person
    b. Dogs cannot work a trail well in extreme heat or rain
    c. PLS cannot be destroyed
    d. Accurate assessment of a partially completed trail is difficult
    e. If used for police protection work, cannot be trusted around people
7. Disadvantages of trailing
    a. Requires the use of an uncontaminated scent article
    b. Time and weather decrease effectiveness
    c. Need multiple dogs on multiple passes to trust trail that simply ends
    d. Problems with drug and Alzheimer's disease subjects have been noted

B. Securing the scent article
1. Examples of good scent articles
2. Decision factors in SAR vs Handler obtaining scent article
3. Common errors

a. Articles has been washed since subject last wore
b. Articles has been handled by a family member or relative
c. Article was never worn or handled by subject
4. How to pick-up scent article
   a. Who picks up scent article
      (1) If already obtained dog team can start work immediately
      (2) Dog Handlers often prefer to pick up article themselves
      (3) Make sure person sent to obtain scent article is well trained in both what a good scent article is and investigation
   b. Do **not** touch. Invert paper or plastic bag. If using plastic bag must be unscented. (Large trash bags often use deodorants).
   c. Label bag, how obtained, who obtained, date, time, location, subjects name, and the items in the bag. Use of sterile gauze pads.

C. General considerations
   1. Secure the PLS. Keep staff, searchers away from exit door if that is the PLS. Don't keep walking in/out door. Turn off vehicles parked close.
   2. Every handler wants a shot at the PLS. Cannot really write a task, dog goes where it wants.
   3. Need walker in excellent shape who can move fast
   4. Walker should be assigned to aid in navigation and handle communications. Walker needs to be able to navigate and plot positions without stopping or slowing the dog down.
   5. GPS units may be used and should be programmed to automatically log position every 5-10 minutes

D. PLS
Objective: Validate PLS, determine direction of travel, find subjects trail, locate subject.
   1. Expanding circle used to locate trail. Dog finds trail and goes.
   2. Value of multiple dogs

E. Specific locations
Objective: Verify clues, locate alternative LKPs, verify sightings.
   1. Physical clues located
   2. Distinctive high probability locations. Former homesite
   3. Containment - Some handlers not trained to perform task

F. The tracking dog dilemma
   1. Several shifts into search
   2. Major uses: sightings, physical clues, surveillance
   3. If dog is dismissed not uncommon to call back if needed

G. Debriefing
   1. Determine validity of scent article. Discuss dogs behavior at start of trail
   2. Discuss dogs behavior if trail lost. Search area round where scent is Lost.
   3. Document the exact route the dog traveled
   4. Discuss any area where the dog seemed to have difficulties with the trail

V. Signcutters/Trackers

A. Basic Definitions
   1. Field Team Signcutter (FTS)
   2. Search and Rescue Signcutter
   3. Search and Rescue Tracker
   4. Term varies in different parts of the country

B. Information FTS, Signcutters, and Trackers may provide
   1. Age of clue/sign/tracks
   2. Direction of travel
   3. Clue/sign/track analysis

C. General considerations
   1. Signcutters/trackers should be members of overhead teams and quick response teams
   2. Signcutters/trackers can be used at any time on a search
   3. Most used as clue/sign/track eliminators or verifiers
   4. Call-out process
   5. Qualifications of person stating that they are a Signcutter/tracker
   6. PLS or LKP must be secured
   7. Concentration span differs for each signcutter/tracker
   8. Set up man-made track traps
   9. Night signcutting can be better than during the day

D. Information to provide Signcutter/Tracker- Briefing
   1. Location of PLS or LKP.
   2. Weather (past, present, and future)
   3. Topo map of search area
   4. What other searchers have been in their assigned search area
   5. Type of footwear (and measurements) reported to be wearing- Often found to be inaccurate
   6. Set of lost subject's other footwear if available
   7. Areas of any man-made track traps
   8. Peculiar gait of subject
   9. Other applicable information from the Lost Person Report

E. Assignment of Walkers to Signcutters and trackers
   1. Dual nature of tasks for non-signcutting/tracking persons
      a. Slow and boring mode
      b. Fast and furious mode
   2. Uses for Walker
      a. Communications
      b. Navigation
      c. Carrying survival and rescue equipment
      d. Interface with SAR world
   3. Assignment usually given to protégé or those with interest in signcutting/tracking

4. Walker must stay behind Signcutter/tracker unless told otherwise
5. Walker must remain quiet unless necessary
6. Other considerations

SEARCH TACTICS

F. Expanding Circle
Objective: Locate clues/signs/tracks and characterize patterns of the subject, follow tracks to subject.
    1. Starting point: PLS, LKP, physical clue
    2. Start at point and go where it goes (see **Fig 3.6**)
    3. If unsuccessful in finding starting point, expand circle out

G. Perimeter search (**Fig 3.30**)
Objective: Locate clues/sign/tracks. Look for possible turnoffs. Determine if subject moved though an area.

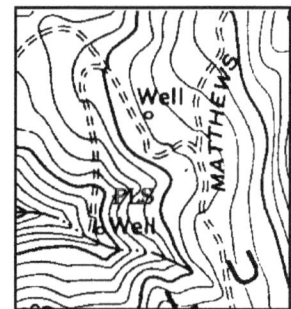

Fig **3.30** *Perimeter Search boundary*

    1. Establish perimeter with features good for tracking
        a. trails
        b. roads
        c. streams, rivers, bodies of water
        d. railroads
    2. Best to signcut in both directions
    3. Similar to a circle search but utilizes roads, trails, stream beds, etc. to signcut an enclosed area surrounding the PLS or LKP.
    4. Size
    5. Cut for sign of passage

H. Field search (internal perimeter)(**Fig 3.31**)
Objective: Determine if tracks of subject in a specific area.

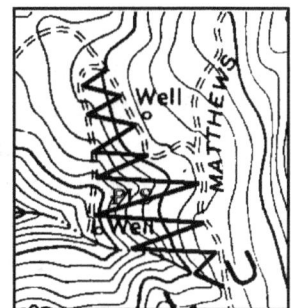

Fig **3.31** *Zig-Zag pattern used to search internal perimeter*

    1. Establish perimeter similar to hasty
    2. Size
    3. Internal Zig-Zag pattern

I. Feature Search
Objective: Follow feature to determine if subject on trail, turned off trail, or turned onto different trail
    1. Starting Point: Typically road, trail, railroad track, creek, gully, etc.
    2. Best to signcut both directions
    3. Following track
    4. Looking for turnoffs
    5. Tracker may jump ahead

J. Other Search tactics include
    1. Field Team Signcutter- as a member of a regular ground or dog search team utilizing any of the previous search tactics
    2. Leap frogging
    3. Cross tracking
    4. Zig-Zag
    5. Construction of track traps
    6. Search for natural track traps
    7. Paired with dog/horse/field teams

K. Debriefing
    1. If clues/sign/track were found
       a. Where
       b. Age
       c. Direction of travel (Could you follow the sign/track?)
       d. Should other resources be sent into the area, should more experienced trackers be sent into the area?
       e. Measurements of the sign/track
          (1) pitch
          (2) stride
             (a) even
             (b) variable or constant
       f. Shoe measurements
       g. Probability of Clue (POC) belonging to the lost subject
       h. Other information form the clue/sign/track
       i. Was clue/sign/track protected/marked? How?
       j. Behaviour pattern determined from gait. (open to interpretation-DANGER.)
       k. Signcutting/tracking techniques used.
    2. Were man-made track traps left? where?
    3. Soil conditions
    4. Weather conditions
    5. Areas not covered
    6. Suggestions for future tasks
    7. Location, route, areas not covered
    8. Tracking techniques used
    9. What discounted and why

L. Staff difficulties with Signcutters/trackers
    1. Feedback may be difficult to measure
       a. Tracking can be esoteric
       b. Ignorance of staff members
    2. Assessments strongly based upon individuals, standardization is much more difficult
       a. What signcutting/tracking standards are being used
       b. Is Signcutter/tracker a member of a SAR tracking organization
       c. Limited number of Signcutters/trackers
       d. Limited availability of Signcutters/trackers

3. Questions staff may want to ask an unknown Signcutter/tracker
   a. Where and when did you get your training?
   b. Do you have a training/search log to present?
   c. How may searches have you been on as a tracking person?
   d. Are you a member of a SAR group or a SAR tracking group?

4. Problems with new (and experienced) signcutting/tracking persons
   a. Shooting from the hip
   b. Want to make nothing into something
   c. Reaching the wrong conclusion based upon limited information
   d. No dirt time or lack of search time
   e. Spends all day looking at a single leaf- or follows ghost tracks
   f. Ego
   g. Wants to give a three hour dissertation on a single sign
   h. Doesn't realize he/she is burned out
   i. Provides interpretation without facts

VI. Horses

A. General considerations
   1. Terrain limitations
      a. Highly mountainous boulder areas
      b. Thick foliage
      c. Swamps, water
      d. Caves

Fig 3.32

   2. Advantages of horses
      a. Terrain advantages
         (1) Trails
         (2) Open fields, open woods
         (3) Fields and woods with moderate brush
      b. Size of task can be much larger
      c. Greater mobility
      d. Rider able to concentrate on clues, while horse concentrates on walking
      e. Higher vantage point (may be disadvantage in some circumstances)
      f. Resource mixing: dogs, trackers
      g. Horse's instinct as prey animal
      h. Endurance
   3. Safety issues
      a. Usually assigned to work in pairs
      b. Helmets, First-Aid supplies for rider and horse
      c. Adequate training to condition horse and rider to unusual conditions
   4. Logistical Concerns
      a. Longer lead time required to prepare horse, transport team, and prepare for task. Feed concerns.
      b. Horse will require area to park trailers and corral horses.
      c. Need to alert horse teams early to give sufficient lead time to prepare

SEARCH TACTICS

B. Trail, linear
   Objective: Cover high probability areas, quickly, using minimal resources.
   1. 2-3 horse units (Horse and rider) per task
   2. Must be clue aware, trained to dismount at trail intersections
   3. Consider transportation of horse for start and end points of tasks
   4. Distance covered will be much greater than walking task. Four to six hours still optimal task time.
   5. Easily combined with several other search tactics (sound, tracking, dogs)
   6. Other factors similar to dog and ground linear tasks

C. Multi-trail/road.
   Objective: Cover a network of potentially unmapped trails/roads within a sector with multiple teams.
   1. 4-8 horse units initially
   2. Two-unit team break off from main group when an unmapped trail or road encountered.
   3. Excellent navigation required to plot unknown trails
   4. Greater logistical needs to determine the start point of the task
   5. Other considerations similar to regular trail task (see page 3-1)

D. Modified Trail/Road Sector
   Objective: Quickly cover the area of highest probability with minimal resources.
   1. 2-3 horses per team
   2. Similar to ground Trail/Road sector tasks (see page 3-2)

E. Containment
   Objective: Keep subject within search area. Locate subject on roads or trails.
   1. Typically used for patrolling roads and trails
   2. Multiple horse teams may be deployed at once
   3. Able to cover more ground than foot team

F. Sector searching
   Objective: Non-Thorough technique to efficiently cover search areas.
   1. In wooded areas grid similar in size to ground sweeps and grids (see page 3-3). Wandering patterns or grid patterns similar to air-scent dog grids.
   2. In open fields typically cover perimeter first, then work internal area using wandering or grid pattern
   3. Important to avoid internal boundaries such as fence lines or walls
   4. Boundary between fields and woods makes excellent sector boundaries
   5. Number of teams and amount of time in sector help to determine target POD

G. Logistical support
   Objective: Provide logistical support to field resources more rapidly than a ground response.
   1. Rapid delivery of personnel for several different reasons
      a. Medic reaching patient.
      b. Removal of some injured patients (sprained ankle) from field
      c. Replacement field team leader or rescue specialist
      d. Law enforcement officials for encounter problems
      e. Tracker to follow-up on reported track
      f. Rapid delivery of equipment into the field

VII. Helicopters

Fig 3.33

A. Landing Zones (LZ); Helispots
   1. Determine LZ
      a. Safe
      b. Close to base
      c. Determine coordinates from map or GPS
   2. Determine Marshal and LZ control
      a. Use local fire department if trained and normal local operating procedure. For a guide to setting up an LZ, review FTM or take local aeromedical course.
      b. Use trained and properly equipped SAR personnel otherwise
      c. Have law enforcement and fire presence if possible
      d. If mixture of helicopter and/or fixed wings need trained air operations

B. Briefing
   1. Give operational objective and history
   2. Give primary and secondary search tasks
   3. Importance of several passes
   4. Clues to look for in search area
   5. Chain of command- who they need to talk to
   6. Map/ Grid system used
   7. Locations of teams in assigned sectors
   8. Signal devices subject or teams may have
   9. Communications

Fig 3.34 *Briefing aircrew*

      a. Call signs/numbers
      b. Frequencies
      c. Availability of radios
      d. Direct communication between aircraft and command post essential
      e. Ideal to also have direct communications between aircraft and field teams
   10. Checking in and out
   11. Weather report if requested
   12. Best to have helicopter land before and after task

C. Scanner (Observer) Considerations
   1. Scanner selection
      a. Many helicopters respond only with a pilot, requiring selection of a scanner(s) to complete the aircrew. Initially viewed as high glory position, with everyone wanting the position
      b. Law enforcement
      c. Ability to read topographic map/handle communications
      d. Scanner background or specific training (scanning/signcutting/tracking)
      e. Individual's SAR experience is not a clear indicator of target detection performance
      f. Airsickness susceptibility
      g. Correctable vision, normal color vision
      h. All scanners should be reminded of the importance of reporting any object that could be the subject or indication of the subject to pilot
   2. Scanner lookout positions on the aircraft
      a. Best positions are forward looking
      b. Positions need to be comfortable
         (1) Windows must be clean
         (2) Consider opening side windows if possible
         (3) Minimize light inside aircraft to decrease window reflections
   3. Reducing scanner fatigue
      a. Ideally shift positions every 30 minutes. Not possible in smaller aircraft
      b. Frequent light snacks and beverage (caffeine acceptable), reasonable amount of conversation
      c. Do **not** use binoculars
   4. Dealing with drops in motivation
      a. Common as search progresses
      b. Keep scanners informed of new clues, developments in search
      c. Competition among crew members as a source of motivation
   5. Effective scanning techniques
      a. Scan patterns should be adjusted to include only areas with a high (relative) POD. Most SAR personnel tend to scan areas that are beyond expected detection range for small search targets. Almost all detection of persons on open water occur within 1 nautical mile of the search aircraft at a relative bearing between 225° and 135°. The distance for inland searching is expected to be much less. Scanners should learn parts of the aircraft that aid in determining effective search angles. Examples include window and door frames.

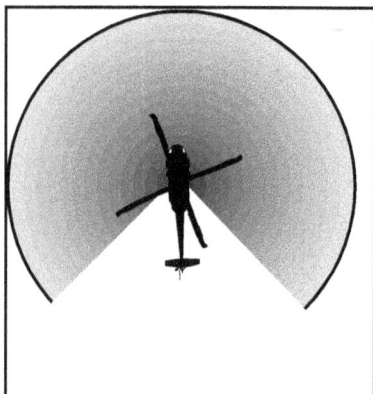

Fig 3.35 *Effective search area from aircraft*

      b. Geometric scanning patterns
         (1) Eye movement should be away from the aircraft to the maximum

detection range, then back toward the aircraft to a point as near under the aircraft as can be comfortably seen. Forward aircraft movement causes the field of view to be moved along.

(2) Eyes should move and pause for each 3° or 4° of lateral and/or vertical distance at a rate that will cover about 10° per second

(3) Sightings typically occur in an area limited by a 5° radius in all directions from the point at which the eyes are focused

(4) Poor scan patterns

    (a) Prolonging eye fixation in a single area

    (b) Allowing eyes to jump around

    (c)   Scanning along structures such as window frames or environmental features within the field of view

(5) When searching in areas with little contrast, eyes tend to focus short. Therefore, scanners should periodically focus on a nearby object within or on the aircraft

c. Mental search image

(1) Look for anything out of the ordinary. Do not look for a complete person or intact aircraft on the ground.

(2) When looking for crashed aircraft detecting broken branches or small pieces of wreckage important

(3) Looking for any differences in outline, shape, color, contrast, texture, and movement

d. Day and Night considerations.

(1) Angle and direction of sun make target viewing different

(2) Targets down-sun: color easily distinguished, best light reflection, less haze

(3) Looking into sun: color difficult to distinguish, silhouette easiest to distinguish, haze most noticeable, heavy glare/eye fatigue. Scanners should wear Polaroid sunglasses

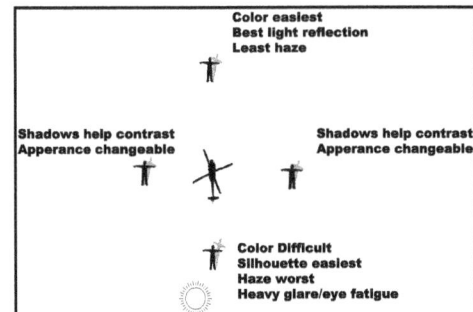

Fig 3.36 *Sun position and target detectability*

(4) Night scanning typically for signals or fires from survivors. Time required for dark adaption (at least 30 minutes). Dark adaption hindered by consumption of cigarettes and alcohol.

6. Airsickness prevention

(1) Take motion sickness medications 30-60 minutes before flight

(2) Consume small amounts of food and water before flight

(3) Ensure adequate ventilation

(4) Keep axis of vision at horizon or distant object

(5) Inform pilot and attempt to vomit into a closed container

D. Measurements of Effectiveness for Helicopters
1. Factors affecting POD include altitude, speed and motion, number of scanners, lookout positions, training of scanners, lighting, duration of search, fatigue, motivation, target detectability, and terrain.
2. Research for looking for lost subjects is incomplete
3. Concept of cumulative POD not valid for air searches in broken foliage

Effectiveness of Helicopter in Desert

# of Passes	POD bright Lighting	POD subdued Lighting
1	31%	65%
2	50%	86%
3	63%	93%
4	71%	96%
5	77%	98%
6	82%	99%
7	85%	"

Table 3.1

POD Aircraft at 500 Feet Searching for a Downed Aircraft

Spacing	Open, Flat	Moderate Cover	Heavy Cover
0.5 mi	35, 75%	20, 50%	10, 30%
1.0 mi	20, 50%	10, 30%	5, 15%
1.5 mi	15, 40%	5, 20%	5, 10%
2.0 mi	10, 30%	5, 15%	5, 10%

Table 3.2 *First POD given for 1 mile visibility, Second POD for 4 mile visibility.*

SEARCH TACTICS

E. Route
Objective: Search along intended route(s) of targets.
1. Follows intended route from point of departure to intended destination (**Fig 3.37**)
2. Multiple passes easily achieved
3. Able to check-out multiple routes if several possible destinations exist
4. Route searches are usually expanded to trackline patterns

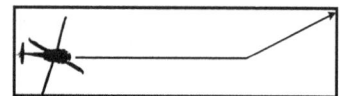

Fig 3.37 *Route search*

F. Trackline
Objective: Rapid search with multiple passes of intended route.
1. Begins with route search
2. Usually first air search pattern since subject located near route
3. Pattern is rapid and gives reasonably thorough coverage of route and area immediately adjacent
4. Two different patterns: Trackline single unit-non-return (**Fig 3.38**) and unit return (**Fig 3.39**).

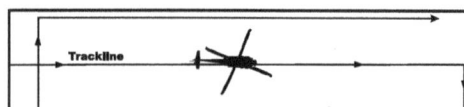

Fig 3.38 *Trackline, Single Unit, Non-Return*

Fig 3.39 *Trackline, Single Unit, Unit return*

G. Parallel Patterns

Objective: Systematic search of a rectangular area with multiple passes.

1. Used to search rectangular or square areas
2. Straight search legs are usually aligned to cardinal headings (N,S,E,W) parallel to long axis of search area
3. Patterns typically used for large, fairly level search area when uniform coverage desired
4. Possible to orient legs parallel to LORAN lines
5. Spacing(s) adjusted depending on desired POD (**Fig 3.40**)

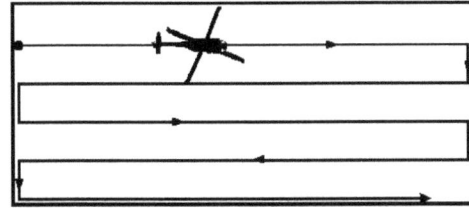

Fig 3.40 *Parallel search pattern*

H. Creeping line grid

Objective: Systematic search of a rectangular area that allows search at one end or adjustments for sun angle.

1. Specialized type of parallel pattern where direction of creep is along major axis (**Fig 3.41**)
2. Used to cover one end of area first
3. Also used to change search direction if sun glare would hinder searching
4. Pattern may be rectangular or elongated

Fig 3.41 *Creeping line pattern*

I. Expanding circle

Objective: Thorough check around center point with an outward curve.

1. Similar to expanding circle pattern used by air-scent dogs and ground teams (see **Fig 3.22**)
2. Overall radius greater than statistical max zone
3. Allows search to be covered beyond the statistical max zone to detect outliers with maximum effectiveness

J. Sector Pattern

Objective: Extensive coverage of center point at different angles and with multiple passes.

1. Similar to expanding circle and square.
2. Used when search area not extensive, concentrated over the central point (often the PLS or LKP)
3. Sectors search area usually do not have a radius > 20 miles, much less for ground searches
4. Easiest to fly, most common in ground searches
5. GPS, LORAN, or visual checkpoint fixes center

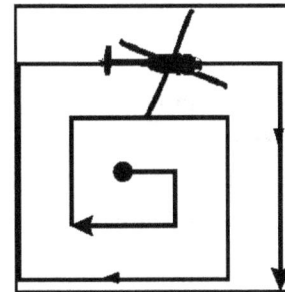

Fig 3.42 *Expanding square*

K. Expanding square

Objective: Thorough check around center point with rectangular legs.

1. Starting point typically similar to expanding circle

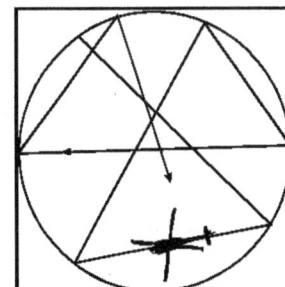

2. Pattern uses straight legs with turns only at corners of each square
3. Requires continuous reprogramming of GPS/LORAN
4. Leg spacing 0.25-0.5 NM in missing person searches

L. Contour

Objective: systematic search of mountainous terrain

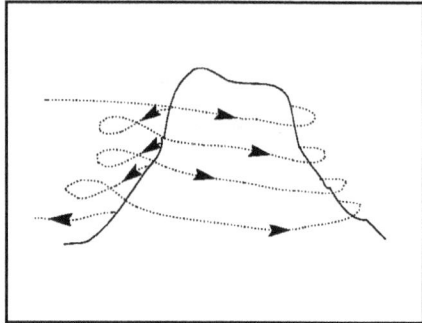

Fig 3.44 *Contour search, may be modified to search ridge line*

1. Used in mountains and hilly terrain
2. Only one aircraft should be assigned to area when conducting contour searches
3. Aircraft starts on the up-wind at the top and works downhill, then returns to the top and works down-wind side
4. As each contour circuit is completed, aircraft typically drops 500 ft in altitude
5. Safety considerations
   a. Crew must be experienced in mountainous terrain
   b. Crew must posses accurate, large-scale topo maps of area 1:24,000
   c. Weather conditions must be good, with good visibility and no turbulence. When winds exceed 30 Kt, down drafts may exceed 2000ft/min. Therefore, aircraft should have a high climb rate. This usually requires reducing the fuel load.
   d. Fatigue factors are high for mountains searches
   e. Accurate briefing and debriefing important

M. Search Segment

Objective: Coverage of a search segment with multiple tactics.

1. Given area to search. Usually combination of sector and parallel tracks
2. Pilots decide on best technique upon arrival on-scene

N. Overview

Objective: Give Command and General Staff an overview of search area.

1. Short flight (15 minutes) to provide IC, General Staff with overview of topology and vegetation in the search area
2. Flight ideally given to staff before their shifts begin
3. Flight should not be coupled to regular search task. Must avoid staff getting stuck in the air for several hours

O. Mapping

Objective: Update topo maps.

1. Scanner must have excellent mapping skills
2. Special equipment may include GPS, LORAN
3. May take Polaroid or 35mm film with 1 hour developing
4. May use videotape to record are
5. Bring copy of topo map with acetate to make corrections

P. Transport

Objective: Transport overhead teams/search teams to search base or to search area.

1. Excellent resource to move ground teams into remote areas or summits reducing ground transport time from base
2. Transport of teams from their home point of origin to search base
3. Weight can be an important consideration, ground teams cannot leave essential equipment behind
4. Must weigh transport need versus expense and safety considerations of any helicopter use

Q. Aeromedical Evacuation

Objective: Immediate evacuation of patient from search area to a medical facility via a medically equipped and staffed aircraft.

1. Aeromedical aircraft may be hospital, private, military, or local government
2. Aeromedical services staffed with flight paramedic, flight nurses, and/or physicians
3. Full-time aeromedical services well equipped
4. Military MAST units have more variable staffing and equipment depending on local protocols
5. Aeromedical service should be alerted once patient is located and a need determined

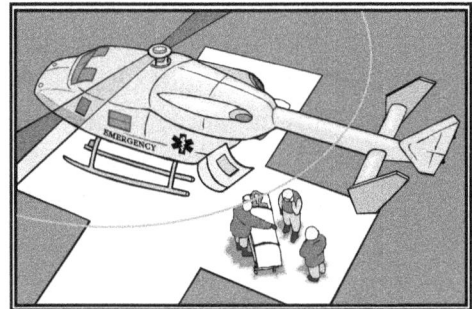

Fig 3.45 *Aeromedical Evacuation*

6. Aeromedical service should be on-scene once patient evacuated to the LZ
7. Many aeromedical services precluded from rescue/hoist/SAR activity by insurance

R. Rescue

Objective: Rescue of search subjects.

1. Helicopter can be used as a platform for hoist, fixed line fly-away, hot loading, single skid landing, and rappel rescues
2. All operations require both proper air and ground crews
3. Excellent communications between Air - Ground and Air - Command
4. On-scene safety officer suggested
5. Make a back-up plan if helicopter unable to perform task

S. Homing Patterns

Objective: Airborne Direction Finding of electronic distress signals.

1. Patterns used to direction find (DF) electronic distress signals
2. Most helicopters do **not** carry DF equipment
3. Coordinated through ground operations
4. After a fix is made, information must be passed onto ground operations

III. Fixed Wings

A. Briefings, scanner selection, scanning patterns, and tactics similar to helicopters

B. If multiple sorties in search area then must be coordinated by experienced and qualified air operations personnel

C. Advantages of fixed wings over helicopters

1. Searching large areas quickly when foliage is not on trees, in open areas, fields, or when subject capable of visual signals

Fig 3.46 *Cessna High Wing*

2. Ideal for communication platform

3. Containment tasks

4. Lower operating costs, much longer endurance and time on task

5. Direction Finding gear

D. Use of private non-SAR pilots not recommended

1. Must use pilots and planes trained to fly slow and low. Special training required for mountainous regions and to properly execute search patterns

2. Private pilots have offered to search in the past. Often more difficult to say no since pilots tend to be better educated than the general public

3. Severe safety hazard if current operations are ongoing

4. Liability is far too high, under all circumstances, if not trained to fly low and slow and search. Margin of error and opportunity for disaster is too high to justify risk

5. Private aircraft not usually DF equipped or carry VHF communications

6. If private pilot insists, offer to coordinate airspace use to avoid two aircraft in same area

E. Tactics

1. Similar to Helicopters: Route, trackline, parallel pattern, creeping, expanding circle, sector, mapping, long-distance transport, and contour

2. Tactics not appropriate for fixed wing: FLIR tasks, short-term transport, rescue, aeromedical from LZ.

SEARCH TACTICS

F. "High-Bird" communications platform

Objective: Provide a communications relay at high altitude to provide communications support to a large area.

1. On large geographic searches or in mountainous terrain may use an orbiting aircraft as a communication platform

2. Aircraft orbits at an altitude that allows communication among all resources

3. Must have correct FCC-type accepted equipment and licenses

4. Limits of task, fuel endurance, human endurance without restrooms

5. Should have a second crew member experienced in manual relays to operate radio

G. Containment

Objective: Keep a continuous patrol perimeter of search area when large.
1. Fixed wing able to keep a continuous patrol of search area
2. Able to continuously search open areas
3. More cost efficient than helicopter operations
4. Able to spot subject if he wanders into open areas

IV. All-Terrain Vehicles (ATV)
A. General features
1. Able to cover trails quickly
2. Carefully evaluate training and survival equipment of operators
    a. Must carry ten essentials/survival/first-aid equipment
    b. Must carry basic mechanic's kit to repair simple problems
    c. Ideally should be trained to same level as searchers
    d. Signcutting/tracking skills must be well developed since footprints will be destroyed after ATV deployment
    e. Helmets must be outfitted to allow two-way radio communications
3. ATV use becoming more common among game wardens and land management agencies
4. Trails must be checked carefully for clues and tracks either by ATV team or before ATV team. Generally, ATV use will destroy any clues. Amber "Trucker" lights should be mounted underneath bumper for tracking at night. Slow speeds essential to good searching
5. Excellent resource for running containment on trails. Must check for clues or tracks at trail intersections and at turnoffs into drainages
6. Should stop periodically, shut off engines, and listen for subject
7. Special logistical concerns include trailers and fuel
8. General volunteers are assigned a low to zero POD. However, they have made finds of hunters, despondents, and Alzheimer's subjects
B. Safety Equipment
1. Helmets, goggles, protective clothing required
2. Survival/first-aid/mechanical gear required
3. Must travel in pairs

SEARCH TACTICS

C. Linear tasks

Objective: Cover high probability area, quickly, using minimal resources.
1. Two ATVs that follows fire road, trails
2. One operator appointed team leader
3. Clue awareness critical, especially at trail junctions and drainage turnoffs
4. Team must be instructed to stop, shut off engines, signal, and listen periodically
5. Make sure team has all required safety equipment
6. Avoid sending unaffiliated ATV operators into woods

D. Transport of equipment or medic

Objective: Rapid transport of equipment or personnel into field.

1. Rapid transport of equipment or medic into the woods
2. Make sure ATV rated for passengers
3. Make sure passenger has helmet and other required safety equipment
4. Avoid excess speed

E. Containment

Objective: Keep subject within search area. Locate subject on road or trails.

1. Patrol of trails and fire roads to maintain containment
2. Destroying tracks less critical
3. Should stop engine, signal, and listen periodically
4. Operator must be properly trained and equipped
5. Containment tasks often require a SAR-trained operator who understands importance of containment
6. Speed not important, must operate ATV carefully

V.   Bicycle Teams

Fig 3.47 *Mountain Bike*

A. General features
   1. Similar to ATV tasks
   2. Speed slower
   3. Ability to detect clues better
   4. Useful for following other mountain bikers lost on trails
   5. Better able to hear subject shouting
B. Safety factors
   1. Helmets required
   2. Survival gear and repair gear required
3. Must travel in pairs
4. Must have experience at traveling off road
5. Must have survival/first-aid training

SEARCH TACTICS

C. Linear tasks

Objective: cover high probability trails quickly, using minimal resources.

1. Pair of riders follows trails or fire roads
2. One operator appointed team leader
3. Clue awareness critical, especially at trail junctions and drainage turnoffs
4. Team must be instructed to stop, signal, and listen periodically
5. Make sure team has all required safety equipment
6. Avoid sending unaffiliated riders into woods

D. Containment

Objective: Keep subject within search area. Locate subject on road or trails.

1. Patrol of trails, fire roads, and roads to maintain containment.

2. Tactics may be used effectively in urban, suburban areas
3. Destroying tracks less critical
4. Should stop, signal, and listen periodically
5. Operator must be properly trained and equipped
6. Containment tasks often require a SAR-trained operator who understands importance of containment

    E. Transport

Objective: Rapid transport of equipment or personnel into field.
1. Rapid transport of equipment or medic into the woods
2. Medic must be trained to ride bike in an off-road situation
3. Make sure passenger has helmet and other required safety equipment
4. Avoid excessive speed

VI. Kayakers
    A. Used in water searches or rescues.
    B. Safety Considerations
        1. Special equipment
            a. PFDs
            b. Helmets
            c. Throw bags
            d. Self-rescue equipment (spray skirt, extra paddle, etc.)
        2. Shore support
            a. Lookouts
            b. Throw bags
        3. Work in pairs
        4. Swiftwater training, both paddling and rescue
        5. Communications
    C. Briefings
        1. Subject description and clothing
        2. Search objective and tactics
        3. Water temperatures during entire search
        4. Water flow and changes during search
        5. Hazards
        6. History of where other subjects have been found
        7. Primary and secondary take-out points
        8. Shoreline support available
        9. Rescue support of Kayakers available

SEARCH TACTICS

    D. Shoreline

Objective: Search along shoreline for subject and clues.
1. Search along shoreline for subject and clues (**Fig 3.48**)
2. One kayaker may serve as a rescue boat

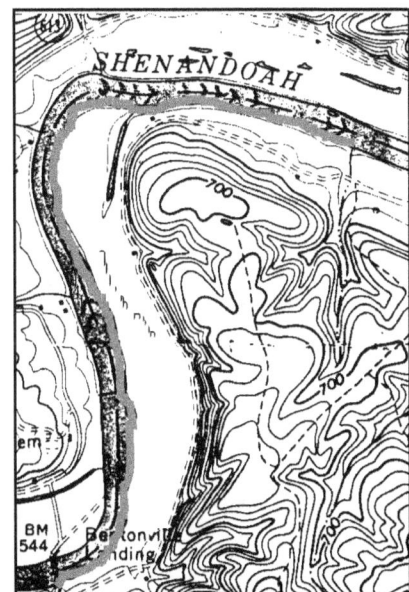

while second kayaker may concentrate on searching in more hazardous areas

> 3. Able to search more hazardous areas
> 4. Kayakers should be instructed to search for clues both above the water and under the water where possible. Particular attention should be paid to snags
> 5. Length of task may be concentrated in one area or for several miles

E. Zig-Zag

Objective: Search along both shorelines and center of river for subject and clues

> 1. Search of both shoreline and center of river (**Fig 3.49**).
> 2. Pattern will be largely influenced by water conditions
> 3. Pattern takes advantage of kayaker's ability to cross or go upstream in swiftwater conditions

F. Creep-line

Objective: Systematic grid of river including both shoreline and center.

> 1. Systematic search of river both along shoreline and center of river (**Fig 3.49**)
> 2. Takes advantage of Kayakers ability to cross river.
> 3. Spacing determines POD
> 4. No statistics exist to determine statistics or formulas for POD. All PODs are subjective
> 5. POD should be expressed for object above water and underwater

G. Rescue

Fig 3.49 *Zig-zag pattern depicted at top of river. Creep line depicted further downstream.*

Objective: Rescue of Subject or endangered "rescuers" in swiftwater.

> 1. Rescue of subjects stranded or trapped by swiftwater. Often includes improperly trained or equipped rescuers
> 2. Safety equipment critical

3. Shoreline support should include lookouts with throwbags, safety lines
4. Helicopter support suggested
5. Must not take unnecessary risks
6. Pairs or teams of Kayakers suggested
7. One or several boats must be assigned a rescue role
8. Proper training is essential
9. Rescue support role may include a back-up if helicopter operations fails, posted downstream as a back-up

UNIT 4: OPERATIONS

I. Introduction
A. Find the subject/get the teams out
B. Safety concerns/always know where your teams are
C. Initially performs the planning function
D. Documentation makes the next shift easier
E. The Big picture
F. The thorn and rose of clues
G. The roller coaster of Operations

II. IC Overview of Operations and information flow

A. General guidelines
   1. Need for tremendous flexibility
   2. Need for staff who can think for themselves and be trusted
   3. Encourage operational staff to make suggestions and question IC decisions in an appropriate manner and setting
   4. Always be prepared to teach
   5. Push gently at the search onset

B. Information structures
   1. Each search has its own unique optimal information flow depending upon a combination of factors
      a. Size and type of search
      b. Experience of staff
      c. Personality of staff
      d. Base and other logistical support structures
      e. Personality and practices of Incident Commander
   2. Might be necessary to try several different information flows before determining the best one

Objectives
☐ Demonstrate the ability to develop and manage a staff and describe when, where, and why various functions should be assigned to which staff positions, including operations and clue analysis.
☐ Describe the internal staff information flow system, including verbal, written and electronic communications, required to insure that information is properly collected, evaluated, disseminated, utilized, and stored throughout the incident.
☐ Describe the differences in deployment of resources in urban, suburban, rural, and wilderness searches.
☐ Demonstrate the ability to complete all necessary operational mission documentation.

The IC makes the bulk of the important operational decisions in the first hour.

When staff demonstrate initiative, organization, and vision, they soon become Incident Commanders.

The Incident Commander must initially push the staff to complete the paperwork.

If every line on every form is properly filled in, does this mean it is a well-run search?

3. Core information tracking
   a. Resource management
      (1) Operations, deputy operations
      (2) Staging area manager
      (3) Resources
         (a) equipment
         (b) personnel
         (c) services
      (4) Tools
         (a) log books
         (b) T-cards
         (c) forms
   b. Task generation
      (1) Operations, Deputy Operations
      (2) Branches, Divisions, Groups
      (3) Plans
      (4) Tactical Plans
   c. Investigations
      (1) Under IC, Plans, Operations
   d. Briefing/Debriefing
      (1) Under Operations, Branches, Divisions, Groups, Sitstat, Plans
      (2) Determination of next task
      (3) Plotting information on maps
      (4) Responsibility for clues
      (5) Responsibility for follow-up
   e. Clue Management
      (1) Can be placed under a variety of positions

Problem
Typical IC thought upon reviewing documentation a week after a search is suspended: "I wish I knew that"
Solution?

C. Information flow verification
   1. Management by wandering and listening
   2. Clue tracking
      a. From radio
      b. From debrief
      c. From outside source
   3. Optimal sequence for a clue
      a. Clue comes into base, plotted on clue log and map
      b. Significant clues brought to the awareness of IC, staff, and investigations. Later field resources become aware
      c. Follow-up action determined
      d. Follow-up action taken
      e. Follow-up action documented on clue log
   4. Task tracking
      a. Task in progress
      b. Task completed
      c. Follow-up progress
      d. Incomplete tasks

III. Operations Section Chief Job Description

A. Collect Initial information:
1. Obtain briefing and report to IC
2. Obtain initial search planning data
   a. Category of subject
   b. Point last seen (PLS) or Last Known Position (LKP)
   c. Circumstances of lost, trip plans (if applicable)
   d. Subject's physical and medical condition, personality traits
   e. Weather
   f. Terrain analysis
3. Obtain searcher data
   a. Name to call (should name be called?)
   b. Physical description
   c. Clothing/shoe/equipment description
   d. Discardables (cigarettes, gum, drinks, ammo, etc.)
   e. Evasive subject
4. Determine resources present and enroute
5. Determine immediate needs for resources and inform IC of these

B. Determine staffing needs and start supervising
1. Determine immediate staffing needs and acquire appropriate personnel
2. Determine if the following functions require personnel assignments: deputy operations, staging area manager, clue supervisor, briefer/debriefer, base radio operator, communications unit leader, division supervisors, branch directors.
3. Supervise and manage operational staff
4. Begin Operations Unit log for major events
   a. Clues found
   b. Important decisions
   c. Resources requested
   d. Finds

The IC should empower his/her staff to further increase staffing if required.

UNIT LOG	1. INCIDENT NAME Shuping	2. DATE PREPARED Dec 17, 1997	3. TIME PREPARED 0600
4. UNIT NAME/DESIGNATORS Operations	5. UNIT LEADER (NAME AND POSITION) Jim Oper (Operations Section Chief)	6. OPERATIONAL PERIOD 0600-2000	
TIME	MAJOR EVENTS		
0620	Appointed Operations Chief, Initial briefing by IC, Sheriff, and investigator		
0633	Briefed staging area manager and deputy operations		
0647	First hasty team (A) deployed		
0652	Scent article secured, locked in deputy Adams squad car		

Fig 4.1 *Example ICS Unit Log*

5. Assist logistics in the placement of base radio system
6. Help select tactical radio frequencies

C. Start initial information management
1. Delegate task of making field maps (OPS staff or SITSTAT):
The method outlined below is only one method of making gridded maps. Many other systems exist. However, it is essential to use one method of gridded maps.

    a. Determine appropriate topographic maps
    b. Mark intersection of Universal Trans Mercator (UTM) grid lines on master
    c. Place ASRC acetate grid aligned with UTM grid mark over map oriented to true north. Write UTM coordinate below ASRC grid numbers
    d. Write on acetate the map letter, magnetic declination, quad name
    e. Adjust copier contrast to illustrate wooded and nonwooded features
    f. Place acetate filters on map if copier cannot achieve above goal
    g. Mark first copy as MASTER and place in master map file
    h. Generally make more copies than anticipated for next operational period

ASRC/UTM Gridded Field Task Map

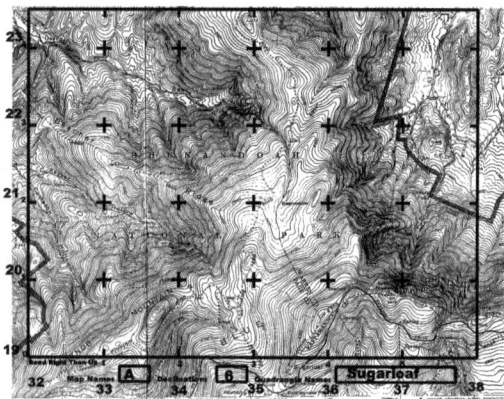

Fig 4.2 *Gridded task map with both UTM and ASRC grid coordinate system. UTM uses two digits while ASRC uses single sequential digit.*

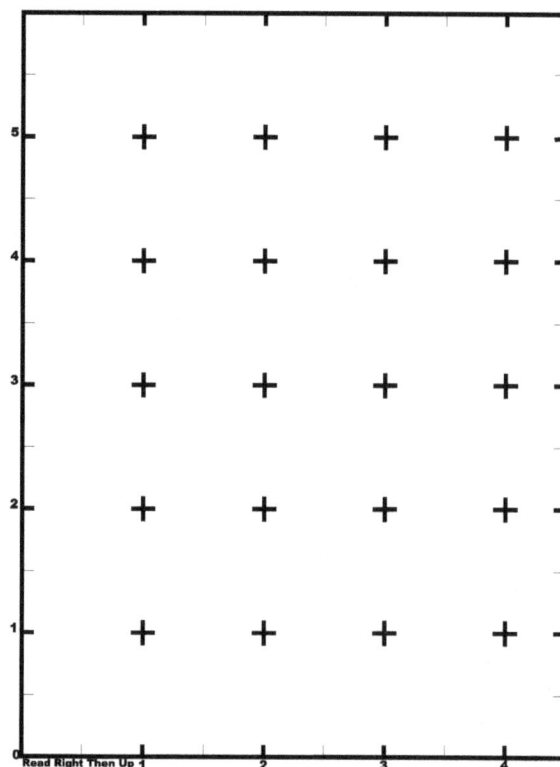

2. To make additional copies of the same field map
    a. Copy directly from master map in folder
    b. Carefully recreate acetate alignment on topo within 25 meters of original

Fig 4.3 *Template for ASRC grid system. Grid based upon 1 km square on a 7.5 minute topo. Grid may be aligned with UTM coordinates as seen in Fig 4.2*

3. Place topographic maps under acetate sheets. Ideally place on map board

with several acetate sheets. Make reference marks so map may easily be realigned with overlays if map removed or jarred.

4. Create Tasks-in-Progress map
   a. Place topo map(s) under acetate sheet or if shortage of original topos and search area is small may use gridded field maps
   b. Mark Point Last Seen (PLS) or Last Known Point (LKP)
   c. Mark Command Post (CP) location
   d. Mark division staging areas if appropriate
   e. Use appropriate colors and symbols (see **Table 4.1**)
   f. In all cases create a legend on the bottom of the map
   g. On larger searches a separate task-in-progress map may be created for dog teams or other resources
   h. Every task placed on map must have its task number to identify the specific team
   i. The task must be placed on the map once the team has been sent
   j. Slash marks with last time of contact may be used to update positions

Task in Progress Map

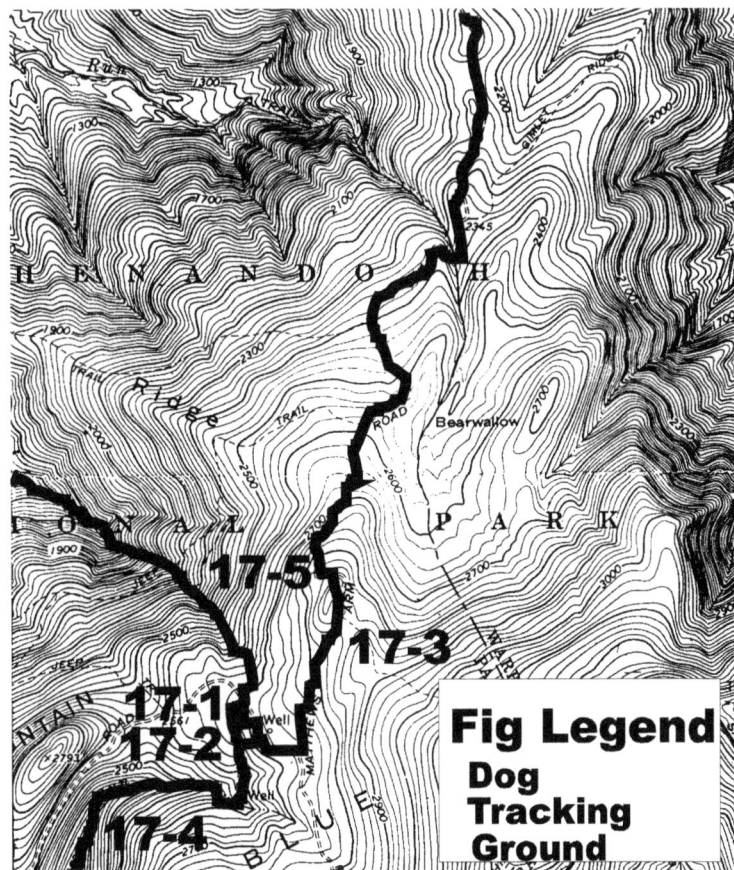

Fig 4.4 *Tasks-in-Progress Map.*

Table 4.1 Suggested Colors for Indicating Search Resources

Black	Section boundaries, barriers	**Purple**	Helicopter/horse
Red	Clues, alerts, PLS, LKP	**Yellow**	Tracking/trailing
Green	Ground search teams	**Blue**	Water
Brown	Air-scent dogs	**Orange**	Special

k. Slash bars or points and ideally, time marks may be used to periodically track the teams' progress

l. When team debriefs or begins a new task, the task is removed from the task in progress status map

5. Start Task Unit log. Complete search and operational period identifying information on the top of the form (see **Appendix E**). Make sure the following information is entered into the log

a. Task number- prefix with date, a hyphen, then task in numerical sequence

b. Team identifier- predetermined designator, letter, or last name. Should follow sequence, same FTL reuses letter

c. FTL's name

d. Team type- (hasty, sweep, grid, containment, air-scent dog, tracking dog, horse, tracker, etc.)

e. Number on team

Task Log	1. Incident name Shuping	2. Date Begun 17 Dec. 1997	3. Operational Period 06:00 - 20:00	4. Page ____1___ of _____ Pages

Task #	Team Identifier	Team Leader	Team Type	#	Task Description	Time Out	Time In	Total Hrs
~~17-1~~	~~A~~	~~Jones~~	~~Hasty~~	~~2~~	~~Hasty around PLS~~	06:47	08:20	1.5
17-2	B	Adams	Tracking Dog	2	Start at PLS	07:10		
17-3	C	Fuller	Tracking	3	Start at PLS	07:23		
17-4	D	Dixon	Hasty	2	Overall run drainage	07:39		
17-5	E	Williams	Air-Scent	2	Sector 5	07:47		
17-6	A	Jones	Hasty	2	Piney run drainage	08:55		

Fig 4.5 *Example Task Log. Once a task is completed it is typically highlighted or marked through with a single line.*

f. Brief task description

g. Dispatch time

6. Create or obtain task folders
    a. Tasks to be done (may be sorted by resource)
    b. Tasks in progress
    c. Tasks completed

Fig 4.6 *Folders used in tracking Task Assignment Form (TAF)*

7. Once a task is dispatched the Task Assignment Form (TAF, see **Fig 4.9**) is placed into the task in progress folder in the order of task numbers

D. **Creation of tasks during first operational period:**
1. Who generates tasks
    a. Tasks may be created at the OPS, deputy OPS, Division, or branch level during the first operational period.
    b. On small searches tasks usually generated by OPS or IC
    c. On medium searches tasks usually generated by OPS, deputy operations, or branch director
    d. On large searches tasks usually generated by deputy operations, branch directors, or division supervisors

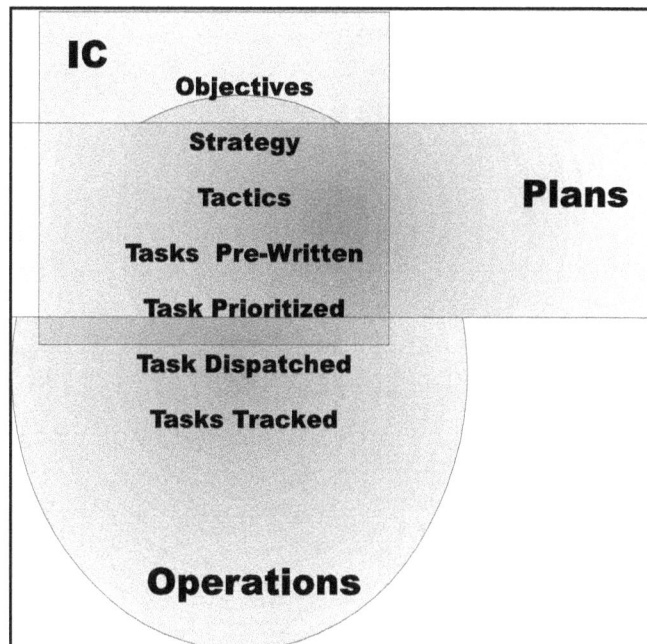

Fig 4.7 *The functional role of the IC, Plans, and Operations will vary depending upon the size and nature of the search.*

2. "Operational" planning of initial tasks
    a. Review theoretical, statistical, subjective, and deductive search areas
    b. Conduct and/or review Mattson consensus if applicable
    c. Consider task ladders priority if applicable or other computer-assisted ranking
    d. Consider types of resources immediately available and incoming
    e. Consider safety, weather, terrain, urgency, subject profile, and other pertinent factors
    f. Consider resource's special capabilities
    g. Attempt to create more tasks than needed for first operational period

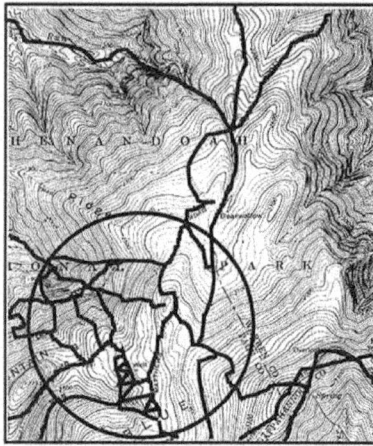

Fig 4.8 *Potential task areas, hasty searches, primary air search area. Tasks can quickly be added to map.*

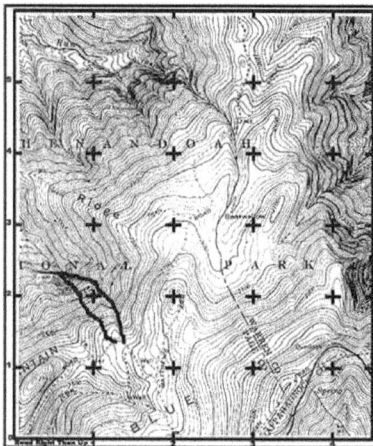

Fig 4.9 *Task area highlighted on gridded task map. Two copies of the task map are made. Once copy goes with the FTL while the second remains in base.*

Fig 4.10 *Pre-Assignment Task Assignment Form printed on NCR paper so that two copies are made.*

3. Create tasks:
   a. *Technique A*- master task idea map
      (1) Place all ideas for potential tasks on a separate acetate-covered map or field map
      (2) Tasks may be color coded for task type (see **Appendix C**)
      (3) Fill in TAF when task dispatched
      (4) Reason to use: Allows ideas to rapidly be placed on paper in a visual format. A second (and often less-trained) person can then use technique B

   b. *Technique B*- double map, TAF pre-assignment
      (1) Technique B may be combined with other techniques as needed
      (2) Take two field maps and with a hi-lighter indicate the area to be searched on both maps
      (3) Fill out the assignment section of the TAF. Complete the important phone numbers, communication, and equipment section if known. Indicate numerical range for size of team. Leave the rest of the TAF blank. Do **not** assign task number at this point.
      (4) Indicate what type of task (hasty, containment, sweep, grid, air-scent dog, etc.)
         (5) Written description of the task must be detailed enough to determine what the task is if the map is accidentally separated from the TAF
         (6) Attach the maps to the TAF and place into the tasks-to-be-done folder. Tasks should be placed into the folder in a priority order. Tasks may be divided into resource types.
         (7) Common mistakes on pre-written TAFs
            (a) Task instructions too detailed, wasting time. Task instructions left blank or simply state "see attached map"

(b) Assigning task identifier and team identifier

   c. *Technique C-* search sectors
     (1) Technique C should be combined with other techniques as needed. Technique differs from Task Planned (technique A) in that only sectors are drawn, may be similar to Mattson map
     (2) Divide the search area into search segments on an acetate- covered map or field map
     (3) Do not specify task type or number until team is dispatched
     (4) Indicate priority order for tasks
     (5) Assign segments to dog, sweep, grid, and wandering teams

   d. *Technique D-* real-time creation
     (1) New tasks may need to be created in order to respond to clues, investigation, or other follow up. These tasks are generally high priority and the task is created and dispatched almost immediately. Technique B may be used except the entire TAF is completed and the team then dispatched.

E. **Multiple individuals creating tasks**
  1. In order to expedite deployment of teams into the fields it may be necessary to have several people simultaneously creating tasks
  2. May be easily assigned to divisions with distinct geographic boundaries
  3. Techniques for smaller searches
    a. Two or more people with equal training may choose a geographic boundary (road, stream, ridge, etc) and pre write tasks in their segment
    b. An individual with more training creates the tasks on a planning map and someone with less training transcribes the task to the task map and TAF
    c. Another individual with less training may pre write "template" TAF with information that will not change during the shift (header information, communication info, safety messages, base phone numbers, and other fixed information)

F. **Creation of tasks during later operational periods**
  1. Review priority placement of TAFs in tasks-to-be-done folder, rearrange if necessary
  2. Continue to create new tasks in order to respond to clues, investigation, previous tasks uncompleted or greatly modified in the field, expansion of the search, covering areas again, or other needed follow up
  3. On small searches OPS or IC continues to generate tasks
  4. On medium searches tasks generated by tactical planners, OPS, deputy OPS, PLOPS, or branch directors

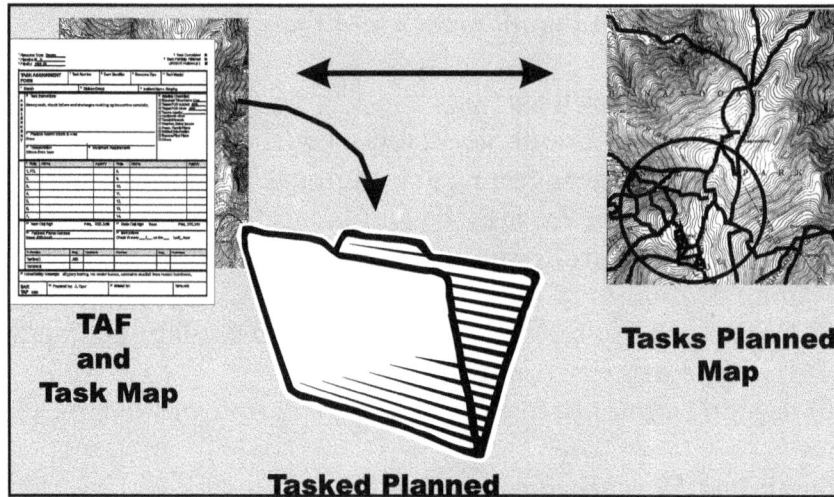

Fig 4.11 *Preplanning documentation for pre-written TAFs. The TAF should be placed in Tasked Planned Folder in priority order. After writing a TAF, the task may be placed on the Tasked Planned Map or the task may be created on the Tasked Planned Map first and then transcribed to the TAF. Planning number on the TAF may also be placed on the planning map for reference. Planning number is NOT the task number.*

5. On large searches tasks generated by tactical planners, deputy OPS, division supervisor, or branch director.

Reflex Tasks

Depending upon the scenario 5-7 tasks are obvious and seldom vary from search area to search area. Linear tasks down the closest drainages and trails, trailing dogs starting at the PLS, hasty of the building, air-scent dogs into the most likely sectors, etc. These tasks allow the IC/OPS to rapidly deploy initial resources.

G. **Initial hasty deployment of teams**
1. This technique is **only** used when initially arriving to a search where rapid deployment of trained SAR FTLs, Dog handlers, and Trackers into the field is required due to the urgency of the situation
2. Place topographic map under acetate
3. Mark the PLS, LKP, and base location on the acetate
4. Visualize the theoretical, statistical, subjective, and deductive search areas
5. Look for terrain features that might attract a subject.
6. With acetate marker (see **Appendix C** for appropriate color) make 3-7 potential tasks (generally linear or high probability dog grids)
7. Brief Field Team Leader

Task Log	1. Incident name Shuping	2. Date Begun 17 Dec. 1997	3. Operational Period 06:00 - 20:00	4. Page __1____ of _____ Pages

Task #	Team Identifier	Team Leader	Team Type	# on Team	Task Description	Time Out	Time In	Total Hours
17-1	A	Jones, Smith	Hasty	2	Hasty around PLS	06:47		
17-2	B	Adams, Strutt	Tracking Dog	2	Start at PLS	07:10		
17-3	C	Fuller, Lunger, Hill	Tracking	3	Start at PLS	07:23		

Fig 4.12 *Task Log filled out for rapid deployment. All team members recorded and task description filled out in more detail.*

8. Record all applicable data on Task Assignment Log (see **Fig 4.13**)
9. Place task number next to task on master map when team dispatched
10. Dispatch team
11. After dispatching teams and when time allows, complete TAF. If TAF not complete when task completed, then fill out TAF during debriefing

H. Methods to expedite teams getting into the field

> **IC/OPS must maintain a sense of Urgency among staff and FTLs.**

1. START ON TIME!
2. The night before brief field team leaders/dog handlers on task
3. Brief FTLs before team members arrive. Increase number of briefers using FTLs, during the beginning of operational period (peak briefing demand). When the demand decreases assign FTLs serving as briefers back into the field. Process may be reversed during the debriefing process
4. Have resources arrive night before, brief, and then sleep
5. Make sure staff in place before field resources

I. Briefing the Field Team Leader

> **Selecting Field Team Leaders**
> **FTLs should be selected based upon previous SAR training and certification. In some circumstances, the IC may choose to field promote. Only the IC may establish parameters for field promotions.**

1. Identify those FTLs currently unassigned. Duty may fall to OPS, staging area manager, or resources situation unit (RESTAT)
2. Select FTL to best match expertise required on the task
3. Option of field promoting must be initially approved by IC. The IC is responsible for establishing standards to be used for field promotions
4. Brief the selected FTL. This function is described under the Briefing/Debriefing Officer. On small searches it is usually carried out by OPS
5. Assemble team. This function is described under staging area manager. On smaller searches it is often carried out by OPS or the FTL

J. Dispatching the Team:

1. Enter task in the Task Log
2. Complete the TAF
   a. Record the task number
   b. Record the team identifier
   c. Type of team (see **Appendix C** for codes)
   d. Time (24-hour)
   e. Dispatcher (person completing the form)

TASK ASSIGNMENT FORM	7. Task Number 17-5	8. Team Identifier Echo	9. Resource Type Air-Scent Dog	10. **Task Map(s)** A

11. Branch	12. Division/Group	13. Incident Name Shuping

ASSIGNMENT

14. Task Instructions

Start at PLS, preceed North along Knob Mountain Road to junction. Enter Keyser Hollow Run drainage. Search both sides of drainage 50 ft until park boundary parking lot. Call for pickup 45 minutes before completion of task.

16. Previous Search Efforts in Area
None

17. Transportation Start on foot at PLS. Call for pickup.	18. Equipment Requirements

15. Briefing Checklist:
☐ Expected Time frame 4 hr
☐ Target POD subject _50%_
☐ Target POD clues 30%
☐ Teams nearby A.C
☐ Applicable clues
☐ Terrain/Hazards
☐ Weather, Safety Issues
☐ Press, Family Plans
☐ Subject Information
☐ Rescue/Find Plans
☐ Others

19. Role	Name	Agency	Role	Name	Agency
1. FTL	Laura Williams	DE	8.		
2. FTM	Frank Kovack	VASRC	9.		
3.			10.		
4.			11.		
5.			12.		
6.			13.		
7.			14.		

20. Team Call Sign Echo	Freq. .160	21. Base Call Sign Command	Freq. .160

22. Pertinent Phone Numbers Base: 555-2435	23. Instructions Check in every ___hr___ on the _____ hour.

24. Function	Freq.	Comments	Function	Freq.	Comments
Tactical I	.160		Medical/Rescue	.205	
Tactical II					

25. Notes/Safety Message: slippery footing, ice under leaves, extensive deadfall after hurricane

SAR TAF 5/96	26. Prepared by: J. Oper	27. Briefed by: Albert Baker	Time out: 07:47

Fig 4.13 *Completed Task Assignment Form (TAF)*

  f. Task map (give letter name)
  g. Division if applicable
  h. Transportation (method and how provided)
  i. Equipment

3. Send the FTL with the TAF to communications, logistics, and/or equipment to complete the relevant sections of the TAF
4. Once the TAF is completed ensure the FTL returns the original with one map of the task (from technique B, see **page 4-8**)
5. Place TAF into Tasks-in-Progress folder
6. Draw task using appropriate color onto the Task-in-Progress map. Indicate the task number
7. Create figure legend at bottom of Task-in-Progress map

8. If appropriate, remove task from Task-in-Planning map (technique A).

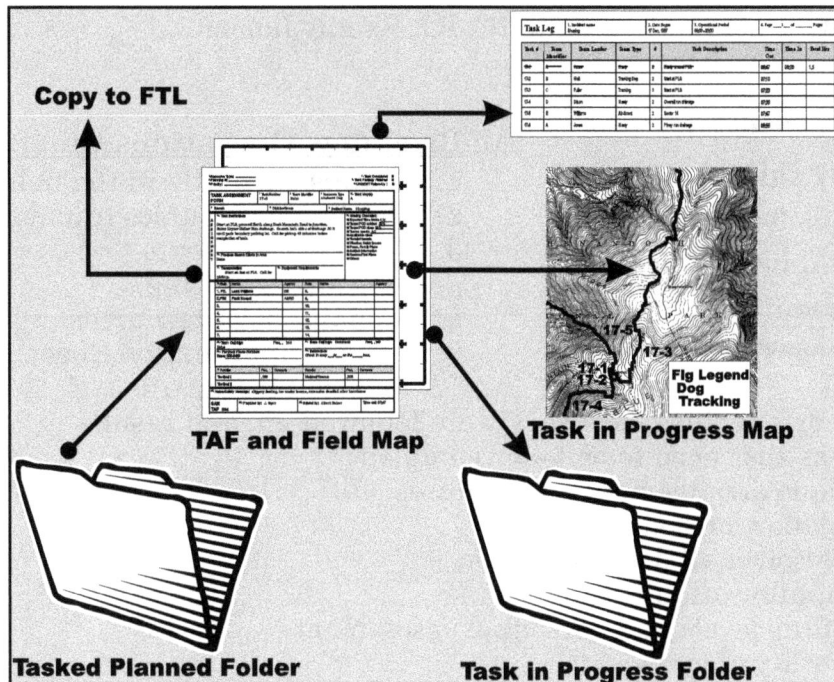

Fig 4.14 *Documentation flow of dispatching a team.*

K. Tracking the Team

1. Team should perform a radio check while still in base and leaving for task to ensure working radio.
2. Team should report upon arrival at search sector and beginning of task
3. Team should report position and status per communication plan
4. Task in progress map may be updated with slash lines and time of last report
5. Communication Unit leader may track teams last check in and take appropriate action when expected check in overdue

L. End of Operational Period Responsibilities

1. At the end of the operational period and/or full calendar day the following information must be summarized
2. Tasks completed folder
   a. Folder labeled with calendar/operational period, day, search name, number
   b. All TAFs arranged by task number in numerical order
   c. Notes explaining any missing TAFs or missing numbers
   d. Task assignment log placed on top of folder
   e. Hard copy (on paper) of task-completed map with appropriate color codes and symbols
   f. Each task on the task-completed map identified by task number
   g. On prolonged searches an additional copy may be done on acetate
3. Review uncompleted tasks or portions of tasks
4. Ensure clue log and clue map up to date
5. Ensure hazards map up to date

6. Brief new staff if requested by new OPS chief
7. Complete appropriate sections of operational period report
8. Provide summary data to IC, IO, Family liaison

> **During every shift change information is lost. Your job is to avoid losing the critical information.**

M. **Brief new Operations Chief:**
1. Teams currently in the field
2. Available resources (numbers and type)
3. Resources enroute
4. Resources leaving
5. Conditions of resources
6. Subject information Sheet
7. Important Hazards
8. Important clues, how they were followed up, and results
9. Tasks that need to be followed up on
10. Unique arrangements, procedures, etc
11. Task flow procedure
12. Location of all documentation
13. Communication capabilities
14. Staffing levels and personnel assessments
15. Accuracy of maps
16. Terrain
17. Weather conditions over the course of the mission

N. **Find responsibilities:**

> **A well-run and smooth rescue requires both**
> ***** **Proper planning**
> **and**
> ***** **Successful implementation**

1. Communications immediately notifies supervisor, OPS, and IC
2. IC notifies AA, SCO of possible find
3. IC may notify AA, Dispatch, and Coordination center of possible find and what actions should be taken at this time
4. Communications ensures clear priority link with find team
5. If report of subject find comes over telephone, verify through AA
6. OPS activates
   a. Medical plan
   b. Evacuation plan
   c. Media plan
   d. Family plan
7. OPS decides based upon preliminary information for other teams to
   a. Remain on task
   b. Hold position
   c. Return to base
   d. Redeployment
8. Once subject status reasonably assured by IC, OPS redeploys teams or starts field withdrawal

9. OPS supervises implementation of medical plan and evacuation plan
10. OPS carries out field withdrawal portion of demobilization plan
11. PSC activates demobilization unit leader if required
12. Communication contacts all teams and verifies instructions
13. IC contacts Dispatch and Coordination center and informs of status

O. **Major Accident Plan**
1. Communication immediately notifies supervisor, OPS, safety, and IC of accident
2. IC notifies AA and SCO
3. OPS activates medical plan and evacuation plan
4. Rescue of known casualty takes priority over an unlocated subject
5. Allocation of resources if multiple known casualties will be made by medical unit leader/specialist using standard wilderness/rural triage principles
6. Resources not required for rescue/evacuation continue search tasks
7. IC notifies Coordination center, information officer, safety officer. IC activates claims unit leader
8. Do **not** release patient's name to media and avoid use of name on radio until next of kin notified
9. IC ensures full investigation of accident conducted
10. IC contacts dispatch or appropriate group dispatch directly
11. IC or designee contacts emergency contact if necessary
12. If possible, IC ensures a team member accompanies any injured team members to hospital
13. Arrange for private briefing to field resources and possible CISD

IV. Briefer/Debriefer (B/D) Function

The Briefing/Debriefing function is responsible for obtaining the TAF and providing the FTL with a complete picture of what is expected of the task and other relevant information. Upon the team's return, the B/D interviews the FTL to obtain pertinent information. The B/D also ensures appropriate documentation occurs.

*   Activation: Whenever OPS, division, or branch director requires assistance dispatching teams

*   Position qualification: GS/IS, experienced FTL, base staff

> **Motivate the team: State the reason you think the subject is in the assigned search area.**

A. Briefing:
1. May be OPS, deputy OPS, operations staff, branch director, division supervisor, or division staff depending upon the size of the search
2. Obtain briefing information from supervisor
3. Obtain Task from supervisor, tactical plans, or tasks-to-be-done folder. Determine purpose of each task

If you don't think the subject <u>might</u> be in a search area, then don't risk a team.

The more experienced the FTL the shorter the briefing.

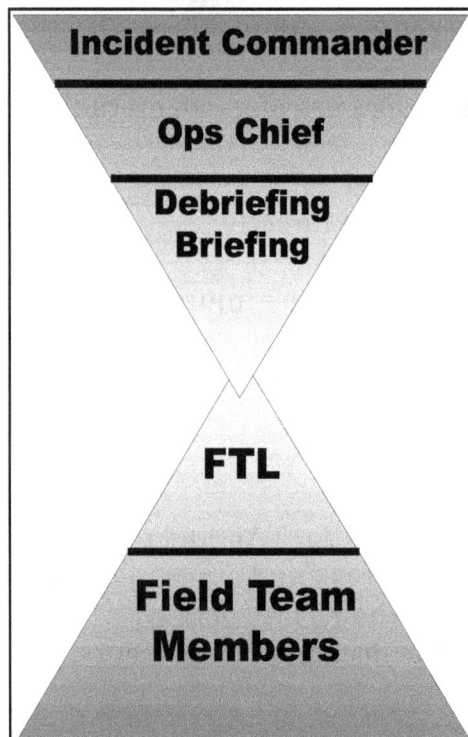

Incident Commander

Ops Chief

Debriefing Briefing

FTL

Field Team Members

Fig 4.15 *Often the only interaction between staff and the field is through information given and collected during the briefing/debriefing process.*

4. Ensure task number, task identifier assigned. Refer to task assignment log if unassigned. Attach appropriate prefix if working in division or branch
5. Ensure team has subject information sheet
6. Ensure team has appropriate field map(s)
7. Brief the team on the following
   a. Task description, rationale for task, how it fits into plan
   b. Allow team to view colored map
   c. Possible terrain and foliage
   d. Required number, equipment, and training of personnel
   e. Target thoroughness of task (POD)
   f. State other teams in or around sector
   g. State if other teams or activity occurred in sector previously
   h. State if follow-up task
   i. State safety hazards, both general and unique to task
   j. Give finding subject procedure
   k. Give medical procedure
   l. Review radio procedures if required
   m. Ensure equipment and radios properly checked out
   n. Review subject information sheet, give applicable information not released to public
   o. Give transportation plans
8. Collect original of TAF with copy of task map
9. Staple material together and place in Task-in-Progress folder or pass on to supervisor
10. Ensure task placed on Task-in-Progress Map with task number shown.

B. Debriefing:
1. Debriefer ideally should be the same person who gave briefing to FTL
2. Remove original of TAF from Task-in-Progress folder
3. Collect TAF, map, unit log, and any clues from FTL
4. Record area actually covered, and how, on the field map
5. Record safety hazards and map changes found. Add these to master map or safety hazards map

6. Record clues found, coordinates, how marked, whether radioed in, and actions taken
7. Ensure clues are entered onto clue log and clue map
8. Confer with OPS or clue supervisor to see if action taken on clues
9. Record areas that need to be checked again on Follow-Up Map and in Follow-Up Log. Pass recommendations to supervisor on need for immediate action
10. Determine terrain encountered
11. Determine morale of team in field. Determine food and water needs
12. Ask and record resource-specific questions
13. Determine readiness of team for next task
14. Determine spread between searchers
15. Determine thoroughness of task (POD) for clues and subject

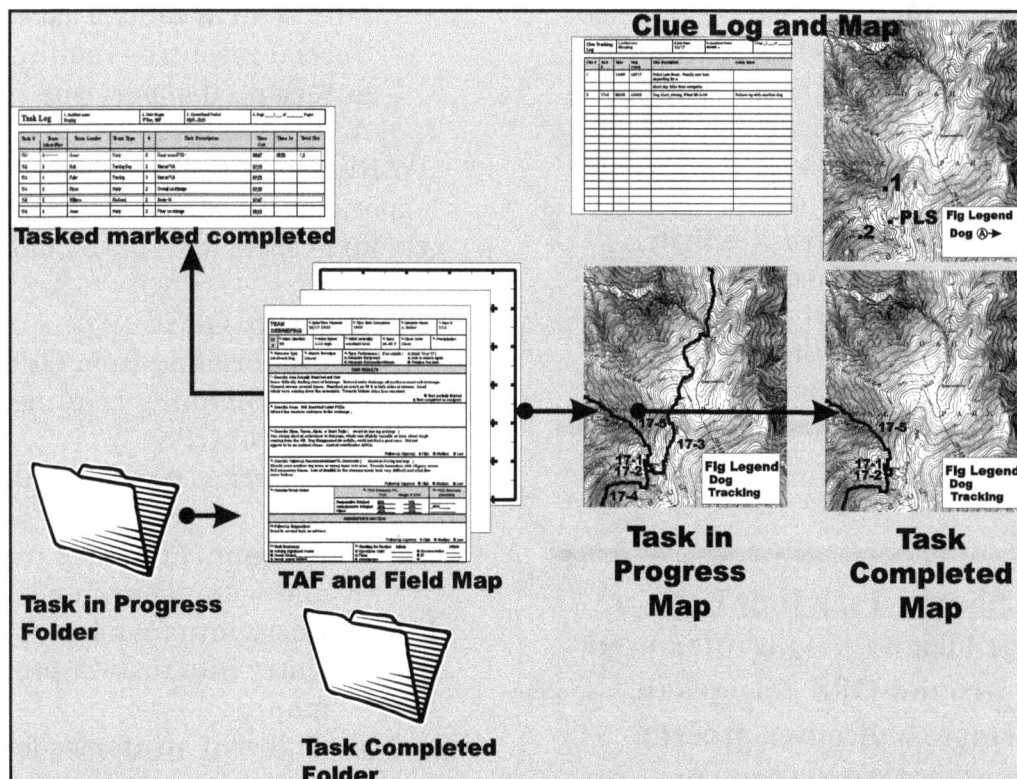

Fig 4.17 *Once a task is completed the TAF is removed from the Task in Progress Folder, the task is highlighted in the task log, and the team is debriefed. The task is then removed from the Task in Progress Map and the actual area covered is marked on the Task Completed map. Clues are marked on the clue map and clue log. The task may be routed to others for review.*

16.

Break search sector into smaller segments if needed for sector searches
17. Determine if FTL has any comments or suggestions for staff
18. Ensure issued equipment returned and checked in
19. Remind FTL of safety procedures (getting sleep, checking for ticks, fluids, etc.)
20. Thank the FTL
21. After relieving the FTL
   a. Collect and staple all paperwork together
   b. Place completed paperwork in Task-Completed folder.
   c. Ensure task removed from Task-in-Progress map
   d.   Ensure task placed on Task-Completed map. Use appropriate color and coding symbols. Identify task with task number.
   e. Ensure task completion entered on Task Unit Log
   f. Keep OPS periodically updated on status of teams in field
   g. Inform OPS, staging area manager, or RESTAT on the availability of FTL and team

V. Branch Directors

Branch directors are responsible for the management of operations, assignment of resources and progress and status of resources appropriate to the branch. Typical ground SAR branches include; Air, ground, water, and law enforcement.

Branch Activation

Air OPS: Multiple aircraft, special coordination required.
Water OPS: Includes both dive and air-scent dog operations on water. Detail of maps different from ground SAR.
Law enforcement: Forensic activity, security, large-scale investigations when investigation is a stated objective.
Ground SAR: May include several divisions, groups on a large operation.

Shared Task Unit Log

A single task unit log is often used if only ground OPS, dog group, and a single helicopter present. Directors and supervisors are located in the same place and share the same form.

---

* Activation:
   OPS span of control exceeded. Special resources require experienced supervisor. Activation by OPS
* Position qualification: IS, experienced FTL, experienced DH, resource specialist, base qualified

1. Report to OPS
2. Obtain briefing from OPS
3. Attend planning meeting at the request of OPS
4. All branches coordinated by OPS for
   a. Assignment of task number and team identifier
   b. Entry into task log
   c. Entry into task-in-progress map
   d. Different strategies for log entries

(1) Single log used by both branch directors

(2) Multiple logs with different system for task designation

(3) Alpha call signs, last name, task number

5. Responsible for managing branch staff

6. Resolve logistical problems reported by subordinates

7. Specialized resources usually assigned to branch director

8. Branch director may keep branch tasks-in-progress map, tasks-in-progress folder, and unit log

9. Responsible for evaluating tasks, modifying priorities, and dispatching tasks generated or provided by OPS, tactical plans

10. Assist OPS and communications in making a branch communication plan

11. Branch director or designee briefs teams

12. Report incoming clues to OPS or clue group for plotting on master clue map and clue log

13. Report hazards to OPS for documentation on master hazard map

14. Branches must closely coordinate all tasks with air-scent dog team group

15. Task-completed activities

    a. Remove task from branch tasks-in-progress map when completed

    b. Place actual task completed on either a master task-completed map (smaller searches) or on a branch task-completed map. Use appropriate colors, symbols, and indicate task number

    c. Place completed TAF and documentation into task-completed folder

    d. Evaluate and follow up on uncompleted portions of task

    e. Maintain clue tracking form (see **Appendix C**)

    f. Maintain unit log (ICS form 214)

Multiple Task Unit Logs

Multiple unit logs are required when branches or divisions are in different geographic locations or when the possibility of task overlap is remote (Air and ground). A letter is typically placed in front of the task number to ensure each task number is unique.

Strategies for Radio Designators

Alpha: (Alpha, Charlie) Easy to understand over radio, reused by same FTL, unique to team, only 24 unless double letter.

Last name: (Koester, Smith) Know who talking to, difficult to understand over radio, same last names common problem.

Numbers: (Unit 1, Team 213) Easy to assign, often confused over the radio, numbers easily transposed.

May use different strategy for each branch or division (dog teams use unit call signs while ground teams use alpha designators).

VI.   Air Operations Branch Director

The Air Operations Director, who is ground based, is responsible for preparing and implementing the air operations portion of the Incident Action Plan. This may range from the coordination of a single helicopter to a complex multi-aircraft air search. The plan must reflect agency restrictions that have an impact on the operational capability or utilization of resources (e.g., night flying, hours per pilot, fuel restrictions). In addition, the Air Operations Director is responsible for providing logistical support to helicopters and fixed wings involved in the incident.

* Activation
    Whenever air resources are deployed.
    Activation by OPS

* Position qualification: Air-operations qualified, if air OPS only responsible for a single helicopter then IS or base qualified.

1. Report to OPS. Obtain briefing from OPS
2. Attend planning meeting at the request of OPS
3. All branches coordinated by OPS for
    a. Assignment of task number and team identifier
    b. Entry into task log
    c. Entry into tasks-in-progress map
4. Responsible for managing branch staff (air attack supervisor, helicopter coordinator, air support supervisor, helispot manager, etc)
5. Organize preliminary air operations
6. Prepare Air Operations Summary Form (ICS 220)
7. Resolve logistical problems reported by subordinates
8. Branch director may keep
    a. Branch task in progress map
    b. Task in progress folder
    c. Unit log
    d. Important to document area the helicopter covers for press and family reasons
9. Responsible for evaluating tasks, modifying priorities, and dispatching tasks generated or provided by OPS, tactical plans
10. Ground-to-Air Coordination
    a. Determine coordination procedures for use by air organization with ground branches, divisions, or groups
    b. Assist OPS and communications in making branch communication plan that allows direct communication between air and ground resources if required. A solid air-to-command frequency is also required.
    c. Coordinate with operations to provide ground resources that are able to quickly respond to any air sightings or clues
    d. Helicopter while flying low may assist or hinder air-scent dog teams
        (1)   Helicopters assist by clearing out scent pools of stagnant air

(2) Helicopters may hinder if operating directly above air-scent dog teams at low altitude above ground level. Effect seen with larger helicopters

11. Supervise all air operations activities associated with the incident

12. Safety related tasks

a. Establish procedures for emergency reassignment of aircraft

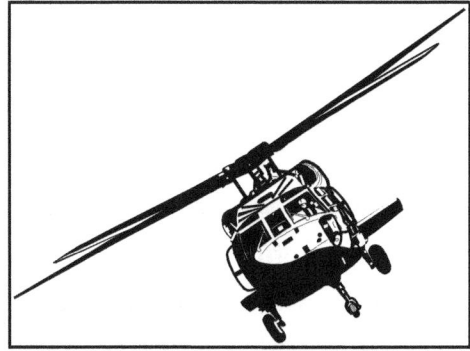

Fig 4.18 *Larger military helicopters may easily interfere with air-scent dogs. They also pose greater challenges for air-ground and air-command communication plans.*

b. Schedule approved flights of non-incident aircraft in the restricted air space area. Inform staff of air traffic situation external to the incident. Consider requests for non-tactical use of incident aircraft. This would include giving staff, family, or press overview flights of the search area.

c. Report hazards, incidents, or accidents to OPS for documentation on master hazard map or other appropriate action

> **Air Operations on Smaller Searches**
> **Because of safety concerns an Air Operations Branch Director may be appointed even on smaller searches. Safety issues of setting up a landing zone, communication with aircraft, and coordination with ground personnel may easily require the full time attention of a staff member.**

d. Work closely with safety officer to ensure all flight parameters are deemed safe

e. Coordinate request for declaration (or cancellation) of restricted air space area (Federal Air Regulation 91.91)

13. Incoming clues reported to OPS or clue branch for plotting on master clue map and clue log

14. Coordinate joint response between air and ground resources

15. Task Completed activities

a. Remove task from branch tasks-in-progress map when completed

b. Place actual task completed on either a master task completed map (smaller searches) or on a branch task completed map. Use appropriate colors, symbols, and indicate task number.

c. Place completed TAF or appropriate form and documentation into task completed folder

d. Evaluate and follow-up on uncompleted portions of task

VII. Division Supervisors

The Division Supervisor is responsible for the management of operations appropriate to the division assignment of resources within the division and reporting on the progress and status of operations.

> **Size and geography are the two most common reasons for starting divisions.**

> **Divisions are similar to smaller type III incidents and require similar trained personnel and logistical support as smaller missions.**

> **Divisions because of geographic separation from the command post require division supervisors who are capable of operating independently of the command and general staff.**

> **It is common to operate one or two divisions out of the command post, thereby reducing logistical and command requirements. Less experienced division supervisors are collocated within the command post.**

> **Divisions are typically created in the second operational period. Commo, logistics, tasks, and personnel must be well established with time to work out problems.**

* Activation:
    OPS span of control exceeded
    Geographic division required
    Activated by OPS

* Position qualification: IC, IS, Division qualified
    1. Obtain briefing from OPS
    2. Report to OPS
    3. Implement assigned portions of IAP
    4. Review assignments and requests or releases resources with OPS' ok
    5. Determine management staffing needs (assistants, logistics, briefing, debriefing, communications, etc.)
    6. Maintain unit log
    7. Report to operations when plan must be modified, additional resources required, extra resources unused, hazardous situations, significant events, and important clues
    8. Generate tasks as needed
    9. Check priority of provided tasks
    10. Maintain division task assignment log. Precede all task numbers by letter of division identifier
    11. Coordinate assignment of team identifier with OPS and communications to avoid overlap
    12. Maintain all division documentation including task logs, tasks-in-progress map, follow-up map, follow-up log, clue map, clue log, safety hazards map, communication log, tasks-completed map, and tasks-completed folder
    13. Brief and debrief teams or delegate function
    14. Evaluate and respond to clues and uncompleted tasks
    15. Resolve logistical problems reported

VIII. Dog Group Supervisor

The Dog Group Supervisor, when activated, is responsible for the management of dog operations, assignment of dog resources, tracking the progress and status of dog resources, and coordinating assignments with the ground operations.

* Activation:
  Activated by OPS.
  Moderate searches with five or more dog teams.
  When special coordination of dog teams required.

* Position qualifications: experienced dog handler, IS, experienced FTL, base qualified

1. Receive briefing from OPS
2. Review previous tasks and alerts
3. Attend planning meeting at the request of OPS
4. Report to OPS when IAP is to be modified, additional resources are needed, surplus resources are available, significant clues are found, or resource assignments need to be modified
5. Secure scent article
6. Evaluate and respond to dog alerts and clues. Report all possible alerts to clue supervisor
7. Maintain dog tasks-in-progress and dog tasks-completed maps
8. Brief and debrief all dog tasks
9. Closely coordinate all tasks with ground OPS branch
10. Coordinate with OPS for assignment of task number, team identifier, and entry into task logs
11. Remove task from branch tasks-in-progress map when completed
12. Place actual tasks completed on either a master tasks-completed map (smaller searches) or on a branch tasks-completed map. Use appropriate colors, symbols, and indicate task number.
13. Place completed TAF and documentation into tasks-completed folder
14. Evaluate and follow up on uncompleted portions of task

Activation of a dog group often allows a dog specialist to command dog teams while still working within the framework of ICS.

On searches with 4-8 dog teams, creation of a dog group is one of the more natural division of resources.

On smaller searches, the dog group and ground group typically share the Task Unit Log and Task in Progress Map. As the search gets larger separate logs and maps may be started.

Fig 4.19 *Careful debriefings of dog handlers are critical when working with unknown teams.*

IX. Clue Group Supervisor

The Clue Supervisor, when activated, is responsible for directing a real-time response to all reported clues. The clue supervisor will document clues, evaluate clues, determine the appropriate response, and act on it.

* Activation:
    Function occurs on all searches once clues reported
    On small searches often combined with OPS, b/d, or communications unit leader. Activated by OPS

Only a fool ignores a clue.

A few clues clearly relate to the lost subject, most clues are undefined, and some clues may misdirect the search.

Clue analysis combines attention to detail, "dogged" investigation, tracking information, and gut instinct derived from experience.

* Position qualifications: IS, investigator, experienced FTL, base qualified

    1. Review previous clues from investigation, clue log, and clue map
    2. Attend planning meeting at the request of OPS
    3. Review Division/group/strike team personnel and incident assignments relating to clue response units (ICS Form 204). Modify lists based on effectiveness of current operations
    4. Report to OPS when: Incident Action Plan is to be modified, additional resources are needed, surplus resources are available, significant clues are found, task or resource assignments need to be modified due to clues
    5. Coordinate clue collection among communications (radio and telephone), debriefing, and investigations
    6. Evaluate all clues reported during operational period and provide appropriate response
    7. Review communications log (radio and telephone) every 2-3 hours to determine if any clues have been missed
    8. If not debriefing teams review completed TAFs every 2-3 hours to determine if any clues have been missed

    9. Supervise resources directly responsible for responding to clues
        a. Investigators
        b. Ground teams
        c. Trackers
        d. Tracking dogs
        e. Searchers/air-scent dogs
        f. Forensic units

10. Maintain operational period clue map and clue-tracking log
    a. Indicate location of clue with appropriate color
    b. Number clue on map to coincide with clue-tracking form
    c. If dog alert indicate wind direction and strength
    d. Initial clue-tracking log once follow-up action taken

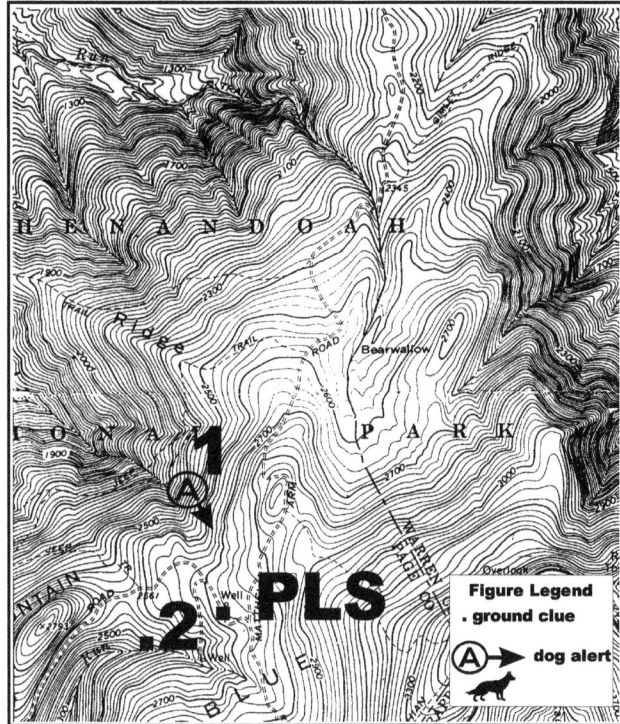

Fig 4.20 *Clue Map. Numbers reference to clue log. Dog alert wind direction arrow follows the dog's nose. Arrows from several alerts will point at the subject. See Fig 4.21.*

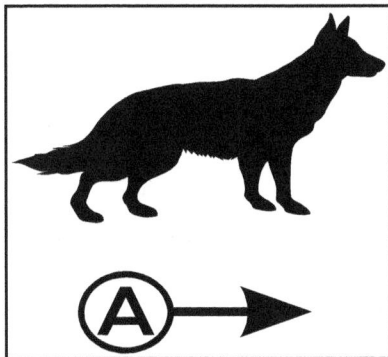

Fig 4.21 *Alert arrow used on clue map points into the direction of the wind. This is identical to the direction the dog's nose points.*

Example Clue Log

Clue Tracking Log	1. Incident name Shuping	2. Date Begun 12/17	3. Operational Period 06:00 -	4. Page __1___ of _____ Pages

Clue #	Task #	Time	Map Coord.	Clue Description	Action Taken	Initial
PLS		14:30	A2717	Point Last Seen. Family saw Ken		
				departing for a short hike from campsite.		*RB*
1	17-5	08:30	A2422	Dog Alert, strong, Wind SE 5-10	Follow- up with another dog	*RB*
2	17-4	08:45	A1305	Foot print, 10 inches long, 3.5 inches	Send tracker	
				wide, heading down the stream. Foot		
				print flagged and covered with plastic.		*RB*

Fig 4.22 *Example Clue log. Clue numbers are referenced to the clue map to give both a text and graphic display of information. Additional lines may be used to elaborate on each clue.*

X. Staging Area Manager

The Staging Area Manager is responsible for managing all activities within a staging area including personnel classification, tracking, and logistics.

* Activation:
    Large searches
    Moderate searches whenever OPS unable to form teams or track resources, may perform functions of RESTAT.
    Activated by OPS

> **An aggressive staging manager can help both field resources and staff in making sure teams don't stand around in base.**

* Position qualification: FTL/FTM with aggressive personality that enjoys details

1. Reports to OPS and receives briefing from OPS
2. Works closely with logistics
3. Establish check-in function on sign-in sheets or computer for
    a. General volunteers
    b. SAR team resources
    c. Vehicle registration
    d. Collect time frame resource is able to commit, include rest and recovery time before departure
4. Orient new arrivals to procedures
5. Respond to requests from OPS, branches, or divisions for resource assignments
6. Inform OPS when resources underutilized
7. Carefully track available IC, IS, and FTLs
8. Work closely with RESTAT, or assume duties. This may include using T-Cards or tracking notebooks (see **Fig 4.23**)
9. Determine projected availabilities of SAR resources using sign-in sheet, availabilities form, computer data base, dispatch information
10. Assist planning in completing ICS form 203 (organizational chart)
11. Establish staging area layout. Determine any needs for equipment, food, shelter, sanitation, parking, security, etc.

PERSONNEL TRACKING FORM

STATUS CODES: A=assigned,   B=available,   C=resting

Name	Affiliation (organization)	Training medical or SAR	Status A B C			Details Team or location	Time/Day of last update	Updated by:
Jim Oper	ASRC	IC III	X			OPS	07:40	*R B*
Sara Adams	SCSO	ODH	X			Bravo	07:40	*R B*
Greg Fuller	SRTA	TK	X			Charlie	07:40	*R B*
Laura Williams	DE	ODH		X		Staging	07:40	*R B*

Fig 4.23 *Personnel tracking form typically kept under acetate. On small searches all names kept on one page. On larger searches similar resources (Dog Handlers, Staff, Field Team Leaders) kept on the same page.*

XI. Field Team Leader

The Field Team Leader (FTL) is responsible for performing and supervising tactical assignments (tasks) given by OPS, divisions, or branches.

*   Activation: whenever resources must be sent into the field

*   Position qualification: FTL, FTM with approval of IC

> **Without the work of field team leaders and team members management would not accomplish anything.**

1. Obtain task briefing from B/D
2. Check out required equipment and radios
3. Obtain personnel
    a. From OPS
    b. From staging area manager
    c. From staging area directly if no staging area manager

4. Return completed copy of TAF or unit log to B/D
5. Brief team on the following
    a. Give name and organization
    b. Give brief description of task
    c. Tell estimated time to task completion
    d. Tell potential hazards and predicted weather for task
    e. Check team members' required equipment
    f. Check for adequate food and water
    g. Determine relevant medical problems of members
    h. Determine if family members/relatives/friend of subject on team
    i. Give transportation details

6. Brief team in the field on the following
    a. Give name and brief introduction
    b. Have team members give name
    c. Describe reason and importance of looking for clues
    d. Describe methods of looking for clues
    e. Give subject description information
    f. Give brief overview of entire search effort
    g. Describe safety hazards and preventative procedures
    h. Describe rest break schedule
    i. Describe assigned task
    j. Describe rationale for task
    k. Describe search mechanics of task
    l. Describe procedures if clue found
    m. Describe procedures if subject found
    n. Describe procedures if team member becomes lost
    o. Describe media procedures
    p. Brief radio operator on established procedures

7. While searching
   a. Carry out assigned task
   b. Monitor all team members for potential safety problems
   c. Provide regular rest breaks
   d. Ensure food and water consumed
   e. Monitor team morale
   f. Constantly track location
   g. Report position coordinates to communications at regular intervals
   h. Place appropriate flagging
   i. Mark any clues found
   j. Report significant clues or events to communications

8. At completion of task debrief the team on the following
   a. Determine area actually covered
   b. Determine thoroughness of area covered (POD)
   c. Review clues found
   d. Review safety hazards located
   e. Review changes in map observed
   f. Encourage team to eat and drink
   g. Encourage team to check for ticks if appropriate
   h. Determine when team or if team ready for next task

9. Report to B/D manager
10. Return issued equipment
11. Turn in TAF/Unit log/map/clues/ and other documentation
12. State when ready to return to field

UNIT **5**: SAR RESOURCES

I. Overview
A. Each resource has its unique strengths and weakness. Diversity should be carefully used to the search's advantage
B. Each group and organization changes over time
C. SAR ultimately comes down to knowing and properly using the strengths of individuals
D. Operational and planning considerations are given
E. Each team must specify in its pre-plan how to contact a wide variety of resources

Objectives
- Demonstrate a complete understanding of SAR resources including important planning considerations.
- Demonstrate an understanding of non-SAR resources including how they are obtained and their potential function in a SAR incident.
- Describe how to effectively and efficiently use non-SAR resources who may offer help at all types of searches.
- Describe how to place a resource order.

II. Federal Resources
A. Active duty military
1. Source of large numbers of personnel
2. Arrive as a true ICS resource
   a. Identified leaders
   b. Categorized capabilities
   c. Sometimes has common communication
3. Usually no formal SAR training
4. Will usually show up without compass, food, water, or raingear unless told
5. Typically arrive in unit transportation
6. Require 6-24 hours to mobilize. Special arrangements or pre-plans may reduce this time
7. Typically arrive in full camouflage fieldgear. Tell unit to bring orange vest or use flagging tape to improve visibility in woods
8. If possible assign SAR trained FTL to accompany team
9. Variable morale
10. Usually used for grid or sweep searches
11. Capable of covering tremendous amount of territory
12. Resource stays until ordered to leave
13. Bases usually have arrangements with local law enforcement to respond directly
14. May provide air resources
15. May require sleeping requirements
16. Method of contact

B. National Park Service

1. Source of Incident Commanders, staff, field team leaders, and field searchers
2. Most fully SAR trained and equipped
3. May respond outside of parks to counties on a boundary with any national park
4. Variable response, not always able to respond outside of park due to shortage of personnel with budget cuts
5. Arrive in POV or in park vehicle
6. If the incident is inside of a park will provide communications, Emergency Medical Services (EMS) and staffing
7. If incident in park can use Volunteers in the Park (VIP) program which provides reimbursement for mileage and lost/broken equipment
8. Some differences between pure ICS and SAR-adapted ICS
9. Method of contact

C. Federal Aviation Agency

1. Provides NTAP (National Tracking Analysis Program) data for last radar plots of missing aircraft
2. Typically gives data to AFRCC (Air Force Rescue Coordination Center)
3. May respond personnel in some situations to help analyze data
4. Method of contact

D. National Transportation Safety Board

1. Investigates transportation fatalities
2. Will evaluate the performance of the search effort in some cases
3. Searchers often will assist collecting evidence for NTSB
4. Method of contact

E. U.S. Coast Guard

1. Provides highly capable helicopter support. Helicopter typically staffed with pilot, co-pilot, and survival-man. Larger helicopters capable of flying in higher winds, IFR, and providing hoist operations.
2. Quickly mobilizes, often within minutes
3. Highly experienced in water rescue
4. Capable of Direction finding a wide range of frequencies
5. Pilots may be uncomfortable flying in mountain settings, especially at night
6. Method of contact

F.   U.S. Forest Service
1.   Source of Field Team Leaders and information about local land
2.   Fully equipped for field work
3.   SAR training usually not highly refined
4.   Small numbers of personnel scattered all over state
5.   Typically requested through local channels
6.   Usually need a day lead time to mobilize
7.   Able to project numbers that will respond
8.   Arrive with some communication capability
9.   Able to provide updates to maps and aerial photographs
10. Able to have bulldozers on flatbeds for rapid deployment (important for making evacuation routes especially with aircraft searches)
11. Method of contact

G.   Federal Emergency Management Agency
1.   Responds resources. Most common SAR-related resources are the Urban SAR teams and Disaster Medical Assistance Teams (DMAT)
2.   Other resources that may be dispatched for active assistance include transportation, communications, public works, fire fighting, information & planning, mass care, resource support, Hazmat, food, and energy assistance
3.   Extensive educational resources
4.   Increasing involvement with the initial phases of disasters
5.   Able to provide extensive logistical support after disasters. Agency has become more proactive in the past several years with a legislative change to the Stanford Act
6.   Disaster operations beyond the scope of the particular course
7.   Major role in disaster assessment

H.   Civil Air Patrol
1.   Most wings can provide SAR-trained ground resources
2.   Mixture of SAR trained and untrained also common
3.   Most arrive SAR trained and with communication
4.   Contain many members less than 18 years old
5.   Not all wings familiar with the Incident Command System
6.   Incident Commander typically called Mission Coordinator
7.   Initial mobilization often slow unless Wing has adopted paging system
8.   Generally possible to obtain projections of responding resources
9.   One of the best resources for protracted incidents/prepared to sleep overnight.
10. Wing Membership

11. Wing Aircraft Resources

12.   Other special resources

13.   Typical equipment found on Wing resources

14.   Flying overhead teams and dog resources

15.   Mobilization time

16.   Contact measures

III.   **State Government Resources:**

A.   State Emergency Operations Center
   1.   Resources able to provide

   2.   Mobile Command Center assets

   3.   Personnel able to provide

   4.   Regional resources

   5.   Mobilization time

   6.   How to contact

B.   State Police
   1.   Law enforcement assets.

   2.   Resources able to provide

   3.   Air resources able to provide

   4.   Mobile Command Center resources

   5.   Regional resources

   6.   Contact method

C. State Forestry
    1.   Resources able to provide

    2.   Fire cache assets

    3.   Personnel able to provide

    4.   Regional resources

    5.   Mobilization time

    6.   How to contact

D. Marine and Inland Fisheries
    1.   Resources able to provide

    2.   Water assets

    3.   Personnel able to provide

    4.   Regional resources

    5.   Mobilization time

    6.   How to contact

E. Game Commission
    1.   Resources able to provide

    2.   Personnel able to provide

    3.   Regional resources

    4.   Mobilization time

    5.   How to contact

F.   National Guard
1.   Total membership
2.   Levels of training
3.   Individual's standard equipment
4.   Team equipment
5.   Method to identify leaders
6.   Management capabilities
7.   Location
8.   Mobilization times
9.   Special skills and capabilities
10.   Ability to project responding resources

G.   Other resources

IV. SAR Resources

A.   Ground Teams
1.   Total membership
2.   Levels of training
3.   Individual's standard equipment
4.   Team equipment
5.   Method to identify leaders
6.   Management capabilities
7.   Location
8.   Mobilization times
9.   Special skills and capabilities
10.   Ability to project responding resources

B.   Air-scent Dog teams
1.   Total membership
2.   Levels of training
3.   Individual's standard equipment
4.   Team equipment
5.   Method to identify leaders
6.   Management capabilities
7.   Location
8.   Mobilization times
9.   Special skills and capabilities
10.   Ability to project responding resources

C. Tracking/Trailing teams
1. Total membership
2. Levels of training
3. Individual's standard equipment
4. Team equipment
5. Method to identify leaders
6. Management capabilities
7. Location
8. Mobilization times
9. Special skills and capabilities
10. Ability to project responding resources

D. Man-tracking teams
1. Total membership
2. Levels of training
3. Individual's standard equipment
4. Team equipment
5. Method to identify leaders
6. Management capabilities
7. Location
8. Mobilization times
9. Special skills and capabilities
10. Ability to project responding resources

E. Mounted Teams
1. Total membership
2. Levels of training
3. Individuals standard equipment
4. Team equipment
5. Method to identify leaders
6. Management capabilities
7. Location
8. Mobilization times
9. Special skills and capabilities
10. Ability to project responding resources

F. Technical rescue/Tactical teams
1. Total membership
2. Levels of training
3. Individual's standard equipment
4. Team equipment
5. Method to identify leaders
6. Management capabilities
7. Location
8. Mobilization times
9. Special skills and capabilities
10. Ability to project responding resources

G. Water Rescue Teams
1.   Total membership
2.   Levels of training
3.   Individual's standard equipment
4.   Team equipment
5.   Method to identify leaders
6.   Management capabilities
7.   Location
8.   Mobilization times
9.   Special skills and capabilities
10. Ability to project responding resources

V. Local Resources

Lost Subjects still need the services of local volunteers.

A. General Comments
1.   Need to change attitudes toward locals resources
2.   Often well motivated, sincere, have valuable local knowledge of subject, area. Several problems possible
3.   Able to contribute significant logistical support once community mobilized. Able to contribute significant field resources. Generally need little mobilization time, highly responsive to request.
4.   Easy to earn the distrust of local resources. Need to request resources earlier in search or local resources will lose interest
5.   Problems with untrained resources include increased liability, less control, increased safety problems if unequipped for conditions
6.   Volunteers should be screened and placed with Field Team Leaders who have been trained to supervise untrained team members

B. Clergy and religious organizations
1.   Able to function as family liaison
2.   Often provide logistical support
3.   Considered source of general community insight and beliefs
4.   Need to establish early contact if plan to use clergy in family plan, especially for finds where subject found dead

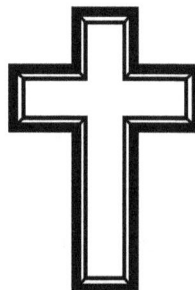

C. Critical Incident Stress Debrief Teams
1.   Some areas have established CISD teams set up for emergency care providers
2.   Other teams are specific for counseling affected family or communities
3.   Training and professionalism of teams can vary widely
4.   Need to have established relationship with team long before a problem develops

    5.   Problems with having CISD team on-scene
    a.   Add additional stress
    b.   Recommendations may interfere with mission objectives
    c.   Lack of understanding with the on-scene role of emergency workers.

D. Coroner
    1.   Need to determine willingness and need to go into the field in case of DOA finds
    2.   Must work into find plan if appropriate
    3.   Work with local law enforcement, need for SAR resources to continue to secure DOA finds until arrival of coroner
    4.   Role of IC to maintain safety and well being of SAR resources may conflict with desires of local law enforcement and/or coroner

E. Child Protective Services
    1.   Source of investigations for insight into family/missing children
    2.   If SAR investigations uncovers possible child abuse/neglect must report to Child protective services. Can route notification through local law enforcement

F. Fire Departments
    1.   Source of search manpower. Fall in the category of organized but untrained SAR volunteers
    2.   Have command structure, some exposure to ICS. Important differences between paid and volunteer squads in utilization on a search
    3.   Communication frequencies often not compatible with frequencies used by SAR resources. Can be used as a second tactical network
    4.   Sometimes able to provide extensive logistical support including
    a.   Transport, scene lights, portable lights, generators, boats
    b.   Firehouses with kitchen/sleeping facilities
    c.   Knowledge of further community logistical support
    d.   Usually has stokes rescue litter

G. Rescue Squads

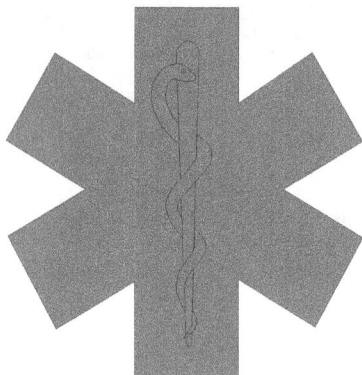

1. Source of search manpower. Fall in the category of organized but untrained SAR volunteers
2. Have command structure, usually little exposure to ICS. Any ICS exposure is for EMS-specific command structures. Important differences between paid and volunteer squads in utilization on a search. Paid squads most likely part of Fire Department
3. Communication frequencies sometimes compatible with frequencies used by SAR resources. Often used as a second tactical network. During initial operations, communications may be run from ambulance
4. Sometimes able to provide extensive logistical support including
    a. Transport, scene lights, portable lights, generators, boats
    b. Squad houses with kitchen/sleeping facilities
    c. Knowledge of further community logistical support
    d. Rescue Squad auxiliaries not as common as in the past. May provide excellent food
    e. Squad may provide resources ranging from medical standby to fully equipped members ready to respond into the backcountry

H. Law enforcement

I.

1. Primary roles include overall responsibility for search, investigation, and security
2. Secondary roles include team escorts when appropriate, extensive knowledge of the area, transport, containment patrols, etc.
3. Best to maintain an on-scene law enforcement at all times to deal with unexpected difficulties
    a. Trespassing
    b. Press arrival
    c. Shots fired
    d. Crime scene finds
4. May be able to provide logistical support, command centers, financial budget, specialized equipment.
5. Tactical team members well prepared to go out into field

VI. Logistical Support Services

A. Red Cross
1. Well-established infrastructure for dealing with disasters/searches
2. Mostly volunteer
3. Many areas have field kitchens capable of providing food for searchers
4. Disaster role includes setting up shelters, providing cots and blankets
5. Need to provide good estimate of number of people requiring meals with sufficient lead time
6. Often need prompting to obtain fresh fruit, bagels, or generally more wholesome food required by searchers for continuous operations

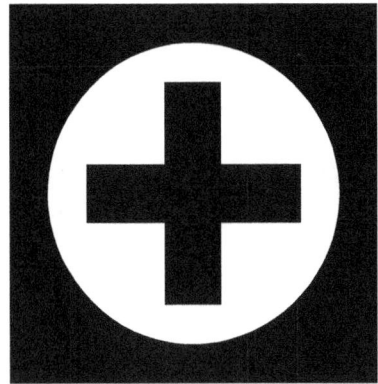

B. Salvation Army
1. Same role as the Red Cross in many rural communities
2. Also tends to have field kitchens capable of serving hot food
3. Less likely than Red Cross to have cots and large numbers of blankets

C. Rescue/Fire/Church Auxiliary Groups
1. Most common role is to provide food and additional fund-raising for local rescue squad/fire/church
2. During searches often provides home cooked meals. Often cooked at home and brought to search. Wide variety of food
3. Able to mobilize quickly

D. Power Companies
1. Provide knowledge of area
2. Can arrange to have fallen trees removed
3. Utility trucks may be able to transport to otherwise inaccessible areas
4. Provide up-to-date information on where poles are located
5. Some power companies have helicopters that may be used for surveys

E. Telephone Companies
1. Able to set up additional telephone lines
2. Always drop in two or more lines when setting up a base
3. Consider special features such as conference calling, call waiting, call roll over, and voice mail features
4. Need to push companies sometimes to provide immediate assistance

F. Local Business
1. Able to offer a wide variety of services
2. Often donate equipment or services
3. Local chain food stores: food
4. Hotels/Motels: rooms
5. Hardware stores: equipment
6. Portable restrooms
7. Copy centers: copies, color flyers
8. Computers, secretarial support
9. Transport Vehicles

VII. Resource Orders

A. Determining resource needs
1. Strive for mixture of resources
2. Determine search area
3. Review search objectives
4. Consider
   a. Search urgency
   b. Political factors
   c. Special skills different teams may offer
   d. Ability of management to handle resources
   e. Need for SAR resources to handle untrained volunteers
   f. Time required for resources to mobilize, travel, and rest
5. Determine types of resources required
6. Determine number of resources required

B. Placing resource orders
1. Write down requested resources before making resource request
2. Be prepared to defend your resource request
3. All Emergency Operations Centers tend to be sparring with resources (personnel, equipment, services) because worried about second incidents, improper use, etc
4. Best to place one resource request per operational period. "Drip" approaches make additional work on everyone and tend to suggest an unorganized base

UNIT **6**: LEGAL AND REVIEW OF APPLICABLE DOCUMENTS

I. Introduction
A. Obsession with legal manners
B. Best to make operational decisions based upon operational needs
C. Companies that strictly follow the advice of lawyers often lose in the big picture
   1. Lawyers
      a. Admit no wrong
      b. Make no operational changes that represents admitting guilt
   2. Tylenol example
D. Probability of lawsuit in EMS field: 1:50,000
E. Factors involved with lawsuit probability
   1. East vs West
   2. Urban vs Rural
   3. First Responder
   4. Civility

Objectives:
☐ Demonstrate an understanding of the laws, policies, procedures, operating instructions, memorandums and agreements which govern SAR operations.
☐ Demonstrate an understanding of certain legal issues related to SAR including; temporary restraining order, trespassing, confidentiality, criminal investigations, management of deceased subjects, restricted airspace, restricting access to various areas, site security and surveillance, maintaining the chain of evidence, use of minors, liability for supplies, equipment, use of SAR personnel for apprehension of criminals and crime scene investigation, discovery of non-incident related illegal activities.
☐ Describe when and how to contact DES and what type of incident information DES requires.

II. Legal issues
A. Medical issues
   1. Temporary Detaining Orders (TDO)
      a. Situation: Family member suffered a mental breakdown and is following teams in the field
      b. Temporary detaining orders allow law enforcement to hold individuals who are a danger to themselves or others for 24 hours, then go before judge. May be longer on weekends
      c. Often used to involuntarily commit someone
      d. Obtain through law enforcement who will contact the magistrate and judge
      e. Person will be taken by law enforcement to a detention facility
      f. Person often given the choice of voluntarily commiting themselves, which means they may then be released by their own decision

Fig 6.1

2. Emergency Care Order (ECO)
   a. Situation: Find a despondent patient who wants to be left alone in the woods. Find a shivering, belligerent, hypothermic patient who refuses care
   b. Allows person who is a danger to themselves to be forcibly brought/detained to Emergency Room/health care facility for intial assesment and emergency medical treatment
   c. Usually it is required to show the patient is in some danger because of physical or mental reasons. May also include the inability to take care of themselves
   d. ECO obtained through law enforcement or Emergency Room who will contact judge/magistrate
   e. Patients will usually cooperate when given the choice of going on their own to the Emergency Room or being forcibly taken
   f. Forcible carry out under the direction of law enforcement
   g. Special care must be taken if using restraints during a long carry-out to avoid cutting off circulation to hands and feet. Most patients can be secured in a stokes basket
   h. ECO are usually valid for four hours once the patient reaches the Emergency Room

3. Confidentiality

   a. All information about the patient's history, condition, and treatment is considered confidential. Confidentiality may be waived by the patient or family
   b. Subject's physical description and circumstances of loss are not confidential. Information entered onto a police report is not confidential. This is the typical method that allows release of information to the public and search teams.
   c. Need to respect subjects confidentiality versus need to help subject
   d. Nothing absolute, different state laws may make it criminal NOT to report information concerning
      (1) suspected abuse
      (2) wounds suggesting violent crime
      (3) suspect sexual assault
   e. Reports should be made to law enforcement or Emergency Room
   f. State code lists exemption from confidentiality requirements to avoid double jeopardy
   g. Law enforcement investigation issues

4. Good Samaritan
   a. State laws developed to provide immunity from civil suits to individuals who render assistance in emergencies. State's laws unique
   b. Provides significant protection in most areas
   c. While government immunity has eroded, good Samaritan protection has increased in most states.

        d. Immunity may need to be proven in court. Doesn't prevent lawsuits

        e. It does discourage many lawyers from taking cases but not private citizens from filing

        f. Good Samaritan does not apply

           (1) If accept payment for services (salary or reimbursement does not negate in some states)

           (2) If gross negligence occurred (needs to be proven in court)

           (3) If care not rendered in good faith

        g. Differences in state laws include

           (1) EMS personnel may be treated differently than public

           (2) Must be an EMT to be given protection

           (3) Specific licenses required to gain protection

           (4) Salaries as a public servant may negate protection

           (5) Need to be well versed in your state's Good Samaritan Law

B. Law Enforcement

   1. Trespassing

      a. May enter property to save life, limb, or property

> **No Trespassing**

      b. May enter any property that is not posted. Must say "No trespassing." "No hunting or fishing" signs do not preclude entry

      c. A property owner without signs may ask you to leave

      d. SAR team members have no special privilages or powers other than good faith or those granted by responsible agent/law enforcement

      e. Field Team Leaders do not like to accept responsiblity for violating law. Proper to shift responsibility to staff or local law enforcement

      f. Options for teams that find "No Trespassing" signs in their sector

        (1) Pre-plan with RA for permission to enter land

        (2) Ask RA/AA to obtain permission on a case-by-case basis.

        (3) Directly ask land owner if identifiable

      g. Most law enforcement officials would not cite a SAR team if a good working relationship exists. Furthermore, most prosecutors would not pursue the case

   2. Criminal investigations

      a. Many searches have a strong criminal element; searches for subjects that are possible murder victims becoming more common

      b. Law enforcement seldom shares all the information with SAR

      c. Trust takes a while to develop

      d. Carefully guard any investigative information given. Only release information truly required for Field Team Leaders to safely complete task. May not release all information to staff. Many SAR members not oriented to maintaining confidentiality.

      e. Law enforcement may be more comfortable only releasing information to one SAR team member

3. Management of deceased subjects
   a. Common occurrence that requires training for all field team leaders
   b. Instructing FTLs, especially non-SAR-trained.
   c. Protection of crime scene major concern. Should avoid placing tarps if possible
   d. Team should have medical protocols for DOA versus starting CPR
   e. Team should consider developing DOA protocols

4. Restricting access to press
   a. Single barrier that keeps out general public may not keep press out
   b. Press has legal right to move closer to scene than the public
   c. Must set up second barrier for purely "safety" reasons
   d. Press has the right to follow in the field
   e. SAR does not have to provide escort or watch press
   f. Best to provide escorts to maintain control. More press details found in Information Officer section

5. Site security and surveillance
   a. SAR members often asked to provide security to aircraft accidents or DOA until proper authorities arrive
   b. Greatest challenge is to keep press out of the area
   c. SAR members have no legal powers unless otherwise deputized or granted powers by law enforcement agency
   d. Many SAR members have no training in security
   e. Proper role is simply to observe site and inform law enforcement if problems develop

6. Maintaining the chain of evidence
   a. In court prosecutor must show any evidence found has been properly handled and accounted for from the time it was discovered until arrival in court
   b. Evidence first seen or found by SAR team members must be properly handed over to law enforcement
   c. Evidence may include scent articles, footprints, and other "typical" SAR clues
   d. Proper documentation critical

7. Use of SAR personnel for apprehension of criminals and crime scene investigation
   a. Many SAR members not trained or equipped to apprehend criminals
   b. Significant liability questions
   c. Many volunteer teams state in SOP team members will not apprehend criminals. Issue should be addressed in team SOPS

8. Discovery of non-incident-related illegal activities
    a. Child abuse, violent wounds must be reported by law
    b. Most common SAR-related find is marijuana plants. Best to report to local law enforcement. Use extreme care, since area may be booby-trapped

POLICE LINE DO NOT CROSS   POLICE LINE DO NOT

9. Crime scene management
    a. Definition: Any location where a crime has been committed or evidence may be found. Procedures also apply to to aircraft accident sites. Crash investigation may uncover possible crime or neglect.
    b. Team should work with law enforcement to develop crime scene SOP since procedures may vary from area to area
    c. Staff must train Field Team Leaders to
        (1) Note and document the condition of the scene. The scene should be flagged or protected at a radius of X feet (radius is best determined by common sense or local policy). Only the medic and possibly FTL should enter the scene to determine if the patient is alive and render care.
        (2) Note the condition and position of the patient. If patient must be moved to render care carefully note and document position and what was moved to extricate patient (important in aircraft accidents). Patient care takes priority over crime scene protection.
        (3) Note fingerprints and footprints. Medics should be wearing gloves and refrain from touching objects as much as possible. FTL should enter and exit the crime scene along the same path.
        (4) Remember and document whatever you touch. Keep rest of team out of crime scene. Important to preserve microscopic evidence. Document information as soon as possible.

C. General issues
    1. Liability
        a. Defined as being held legally responsible
        b. Field Team Leaders should shift as much liability to staff as possible
        c. SAR teams should shift liability (overall responsibility for search) to local law enforcement agency. Written delegation of Authority should always state the overall responsibility rests with RA/AA
        d. SAR teams may also shift liability to State/Federal Agencies through written Memorandum of Understandings
        e. Government agencies often able to provide liability protection (provide lawyers and liability insurance)

    2. Liability for supplies

a. SAR teams often request resources (personnel, equipment, services)
b. Resource requests should be routed through local law enforcement/state agencies/federal coordination centers in order to shift liability
c. Clearly written SOP and MOU help avoid confusion

3. Restricted airspace (Temporary Flight Restrictions)
    a. Temporary flight restrictions (TFRs) restrict aircraft operations over the site of disasters or other areas where rescue or relief activities taking place
    b. FAA authorizes TFR by area manager at Air Route Traffic Control Center (AFTCC) having jurisdiction over the area. If the purpose of the TFR is to prevent an unsafe congestion of sightseeing aircraft above an incident then need approval of regional air traffic division manager.

    c. TFR may be requested by military command headquarters, regional directors of Office of Emergency Planning/Department of Emergency Services, state governors, or similar authority. Best to route request through agency that makes request on regular basis.
    d. When requesting a TFR the FAA needs
        (1) Name and organization making the request, contact number and agency name responsible for on-scene activities
        (2) Brief description of the situation
        (3) Estimated duration of restriction
        (4) Description of area by reference to prominent geographical feature shown on an aeronautical map or VOR/DME fix
        (5) Description of activity posing hazard to persons in the air
        (6) Description of hazard that would be worsened by low-flying aircraft or rotor wash
        (7) Contact point or radio frequency for handling news media request to operate at altitudes used by relief aircraft
    e. TFR cannot be used to exclude media simply for the sake of keeping the media out of the area
    f. TFR are issued by a Notice to Air Men (NOTAM). If pilots do not check recent NOTAM before take off no guarantee TFR will be followed. Difficulty of immediate enforcement of TFR
    g. Not as useful a tool as some search managers are led to believe

4. Use of minors in SAR incidents
    a. ESAR/CAP are major SAR resources with a large percentage of minors
    b. Larger organizations typically provide liability and accident insurance
    c. Organizations may have special requirements for sleeping

arrangements and supervision if coed. Best to let agency
representative work out details

    d. The roles minors play may be limited in cases of

        (1) Security tasks

        (2) When lost subject poses safety risk (suicidal with weapon,
psychotic, history of violence, criminal)

        (3) Body recoveries

III. Legal documents

  A. General comments

  B. National SAR Plan

    1. Applicability

    2. Objective: Integration into a cooperative network throughout the United
States available; US SAR facilities which can be coordinated in any one
area by a single federal agency

    3. Scope: Provides direction to signatory federal agencies. States retain
established SAR responsibilities within their boundaries for
incidents primarily local or intrastate in character

    4. THE PLAN

      a. SAR Areas

        (1) Inland Area- Continental US except Alaska, waters
under the jurisdiction of the US

        (2) Maritime Area- Waters subject to the jurisdiction of
the US; Hawaii; portions of Alaska, high seas and
those commonwealths, territories and possessions of
the US

        (3) Overseas Area. The inland area of Alaska and all
portions of the globe not included within the inland
area or maritime area

      b. SAR Network: Calls for agreements for cooperation

      c. SAR Operations

        (1) SAR Coordinators should develop plans if military assistance
unavailable because of changes in military missions

        (2) SAR Coordinators may request assistance from Federal agencies
having SAR capabilities

        (3) SAR Coordinators should coordinate and direct operations of SAR
resources committed to any SAR mission. On-scene coordination

may be delegated to any appropriate unit participating in a
particular incident

(4) No part of SAR plan should be construed as an obstruction to
prompt and effective action by any agency or individual to relieve
distress whenever and wherever found

(5) National SAR plan does not include SAR for military undersea
rescue, special operations of the Armed Forces, rescue in outer
space, enemy attack, insurrections, civil disturbances, and public
disasters or equivalent emergencies that disrupt the usual process
of government

(6) Federal Government does not compel state, local, or private
agencies to conform to a national SAR plan

C. OSHA issues

1. OSHA regulations do not apply to
   a. States that have adapted their own equivalent plans.
   b. Volunteers (Wholly volunteer organizations in which members receive
      no monetary or other compensation for their services do not fall under
      OSHA purview)
   c. Employees of State and local governments. (Section 3 (5) of OSH Act)

2. Rescue regulations exist for the following areas: confined space, hazardous
   waste and emergency response situations, working on or near unguarded
   energized power conductors, trenches and excavations, diving operations,
   and working around or over water

3. Infection Control Program
   a. Identification of those likely to come in contact with HIV or Hepatitis
   b. Training of personnel
   c. Hepatitis B vaccine provided
   d. Record keeping of vaccine
   e. Body substance isolation protection provided

D. Memorandums of Understanding (MOUs)

1. Typical examples
   a. Between State and Federal Agencies (MOU with AFRCC)
   b. Between Local Government and State
   c. Between SAR agency and local/state government
2. Review applicable MOUs

IV. Applicable Documents

A. EOC SOPS
B. IC SOPS
C. Statewide Preplans
D. Team SOPs

UNIT 7: GROUND OPERATIONS FOR MISSING AIRCRAFT

I. Plans, Agreements, and Policies Governing Air SAR

A. National SAR Plan

B. State-Air Force Agreement

C. State-CAP Agreement

D. Role of State Agencies

II. The Aeronautical Charts
A. Sectional/State chart comparison
B. IFR low-altitude charts
C. Instument Approach Procedure Charts.

Objectives
☐ Briefly review plans and agreements governing air SAR.
☐ Understand basic tools used in flight operations, including flight plans, aeronautical charts, and airport directories.
☐ Discuss aircraft communications and navigational systems.
☐ Review types of SAR aircraft, including their use and limitations.
☐ Discuss crash statistics and behavior.
☐ Discuss air search planning and techniques.
☐ Discuss Ground Operational planning and techniques.
☐ Discuss Scene control.
☐ Discuss elements of clue management

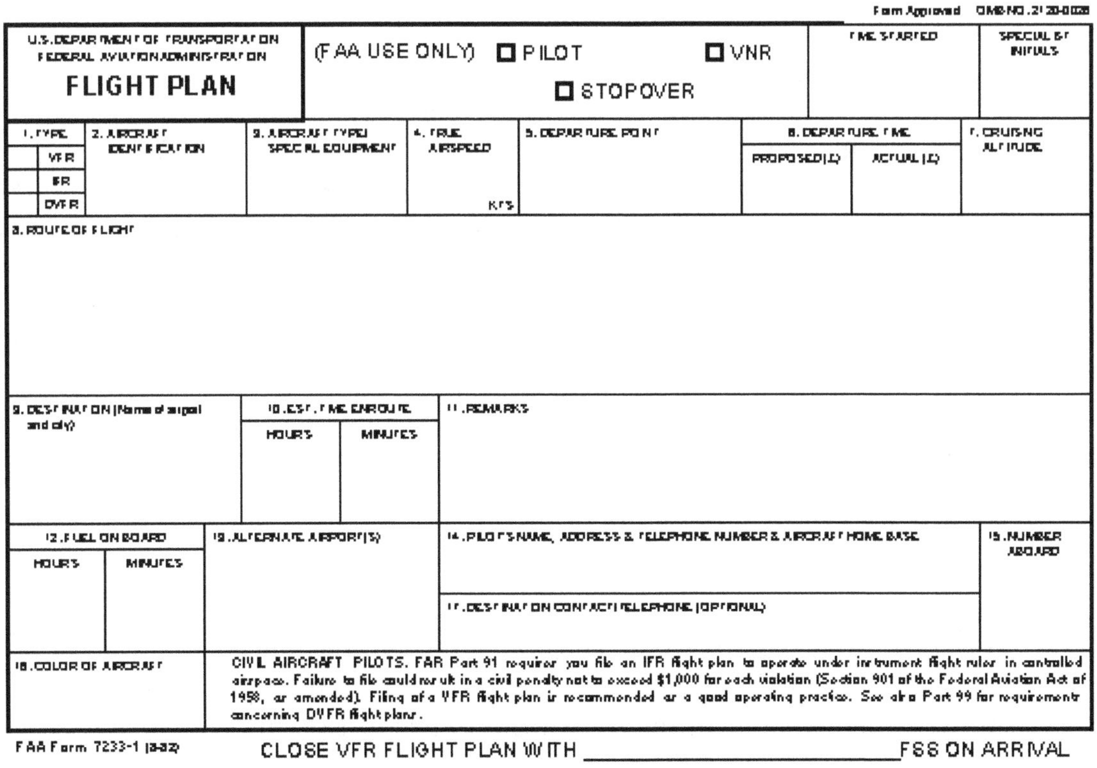

Fig 7.1 *Blank flight plan. A flight plan is not required to be filled out for VFR flights*

III.   Flight planning, navigation, and aircraft systems.
A. Flight plans
B. VOR/DME, VORTAC(VHF Omnidirectional Range Station/Distance Measuring Equipment, VHF Omnidirectional Range Station Tactical Air Navigation)
   1. Navigation system
   2. Pilot flying IFR (Instrument Flight Rules) depend upon heavily
   3. Slowly being replaced by GPS (Global Positioning Systems) navigation
   4. Many crashes along common VOR routes or below VOR transmitters if placed on mountains
   5. Some pilots will give position based upon intersection of VOR radials

Fig 7.2 *VOR as it appears on aeronautical map.*

C. LORAN (Long Range Aid to Navigation)
   1. Now defunct, Signal no longer broadcast

D. Global Positioning Satellites (GPS) Navigation
   1. Based upon a network of satellites that transmits position and time
   2. Replacing both VOR/DME and LORAN within soon
   3. Reads LAT/LONG in degrees, minutes, and **thousandths** of minutes
   4. Accurate to within 3 meters, usually much better
   5. Computer can convert to UTM coordinate system

E. Transponder
   1. Transmits assigned code to Identify Aircraft when swept by radar
   2. 1200 VFR, 7700 Mayday, IFR/commercial flights given a discrete code
   3. Many crashes occur with pilots flying VFR (Visual Flight Rules) in poor weather. They will typically set the transponder to 1200 which is shared by all planes flying VFR. This will make locating the plane using NTAP (radar information) more difficult. Aircraft flying IFR are assigned discrete codes unique to the plane or flight.
   4. MODE 3 includes altitude encoding. Depends on plane's altimeter. Mode 3 must be turned "on" by the pilot if the plane is equipped.

F. Aircraft Radio Systems
   1. Operate around 118.000 - 135.975 MHZ AM
   2. ELT 121.5 Military ELT 242 SAR 123.1
   3. Other frequencies- HF, UHF, CAP
   4. 123.1 common to all aircraft
   5. Programmable radios increase communication capability
   6. Call signs: Last three digits of registration number, "CAP Flight 518"
   7. Ability and considerations in talking to ground teams

IV. Types of Aircraft Used in SAR

A. Light fixed wing
  1. C-150, C-172, C-182 (high wing) most popular
  2. Limitations, 500'-1000' AGL 70+ MPH
  3. Available from CAP, State aviation, Local government, private

Fig 7.3 *High Wing Cessna*

B. Helicopters
  1. Light, heavy
  2. Short endurance, small area, costly to operate
  3. Can hover and operate at low altitudes
  4. Forward Looking Infrared (FLIR)

C. Ultralights
  1. Advantage of able to fly low and slow
  2. Advantage of able to take off and land from small fields

Fig 7.4 *Military Helicopter*

  3. Significant safety problems that must be addressed
     a. Opportunity for accident extremely high
     b. CAP prohibits use of ultralights since risk of accident outweighs benefits
     c. Operators not required to be pilots
     d. Heavy winds or drafts in mountainous areas preclude use
     e. Aircraft has little to no crash worthiness/safety equipment
     f. Pilots must be trained in SAR operations
     g. Hands-free communications for air-command and air-ground teams required
     h. Aircraft must be capable of carrying trained observer to ensure pilot remains focused on safe flight operations
     i. Personal safety equipment required

D. Special Hazards of Search Air Operations
  1. Wind and turbulence
  2. Slow speed and low altitude
  3. Mountainous terrain
  4. Multiple aircraft in grid sector
  5. Helicopter pilots unable to see ground wires from the air

V. Crash Statistics and Pilot Behavior

A. General aviation accidents
  1. 90% pilot error
     a. 40% penetrating known adverse weather
     b. 30% conducting unwarranted, low-level maneuvers
     c. 15% Alcohol
  2. Remaining Causes
     a. Carbon monoxide poisoning

     b. In-flight heart attack
     c. In-flight hypoxia
     d. Mechanical problems

B. Typical Crash Scenarios related to SAR
    1. Scud running
        a. Scenario
        b. Result
            (1) Flight into adverse weather with no out
                (a) exceed structural limits of aircraft
                (b) loss of control
            (2) Spatial disorientation
            (3) Flight into mountain
    2. Takeoff/Landing
        a. Pilot fails to obtain or maintain flying speed
        b. Inadequate pre-flight preparation
C. Result of Spatial disorientation
    1. VFR pilot gets into clouds and loses sight of ground
    2. If pilot flies by the "seat of their pants" (versus looking at instruments)
        within 30 seconds pilots begins to "sense" they are turning left
    3. Disoriented pilot then turns right to correct
    4. This places plane in a constant right spiral downward

D. FAA Survivability Statistics
    1. 50% of passengers are alive 8 hours after the crash.
    2. If injured survival decreases 80% after 24 hours
    3. Only 10% of passengers are alive 48 hours after crash.
    4. Percentage of survivors found as a function of time

% of Survivors located	52%	68%	75%
Time (hours)	12	24	48

E. Missing Aircraft Survivability Statistics
    1. Aircraft crash in more remote areas
    2. 90% of occupants killed upon impact
    3. 6% of occupants injured upon impact
    4. Of the 10% who survive impact 60-80% will not survive first 24 hours
        after the crash without rescue
F. Search Area Statistics
    1. Track analysis statistics
        a. Modified Offset and Track Variable (MOTV)
            (1) Older system of defining search area based upon crash statistics.
                Broke search area into three overlapping areas Newer study has
                resulted in a revised method
            (2) New Two-Area Method (NTAM)

Method requires a Last Known Position (LKP) and a known destination. A trackline is then drawn between the two.

(a) Where enroute turning points have been specified in a missing aircraft's flight plan, the areas should be adjusted so the boundaries of each area are in an arc, centered on the turn

(b) Area one is a rectangle with an offset of 10nm around the trackline. Area two is a rectangle starting at the LKP with an

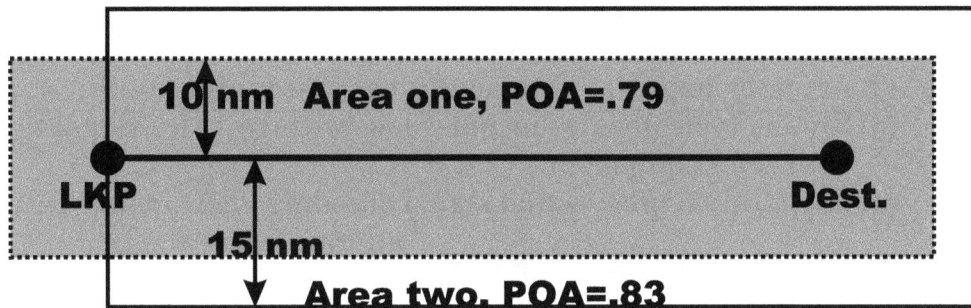

Fig 7.5 *NTAM search area*

offset of 15nm around the trackline

(3) Other statistics

(a) 50% of crashes found on the centerline

(b) 62% of all crashes are clustered within 5nm of trackline

(c) Cluster of crashes related to take-off/landing

(d) Navigation errors, mechanical problems, and weather problems are spread uniformly along track

(e) Cases that occur in the last half of the track are a result of pilots "pressing-on-regardless" under adverse flying conditions

b. Mountainous VFR (MVFR)

(1) In mountainous regions crash sites tended to cluster close to VFR routes through the mountains and were spread fairly uniformly along them

c. Trackline statistics are not detailed enough to start a ground search

2. SARSAT ELT statistics

a. Characteristics of 121.5 and 243 generation satellites

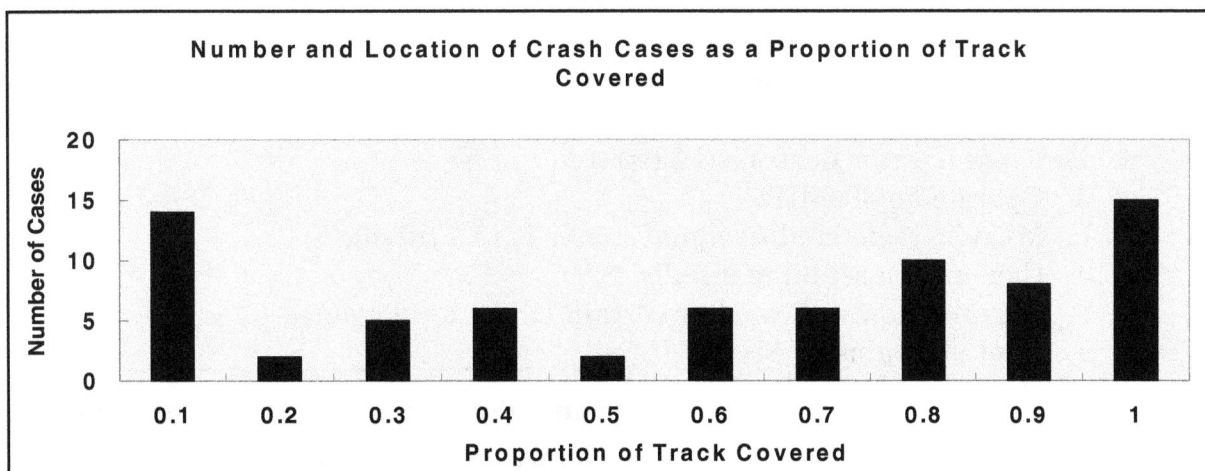

Fig 7.6 *Distribution of crashes along flight track. Clustering occurs at take-off and landing. Crashes along second half of the flight track related to pilots flying into adverse weather and fuel exhaustion.*

(1) Accuracy of fix 5-10 miles
(2) Accuracy improves with additional passes of satellite
(3) Satellite serves as a repeater passing on ELT information to ground station only when satellite is in range of the ELT signal
(4) Older generation ELTs prone to 98% false alarm rate

  b. 406 MHZ ELT System
(1) Accuracy of fix 1-3 miles
(2) Accuracy improves with additional passes of satellite
(3) System typically also includes low wattage 121.5 signal to allow DF with older equipment
(4) Registration process and signal encoding allow identification of the aircraft from its signal
(5) SARSAT package will be placed on geosynchronous satellites for immediate alert (polar orbit satellite still required for fix)
(6) Future 406 system will include GPS technology to allow transmission to satellite of position and aircraft registration

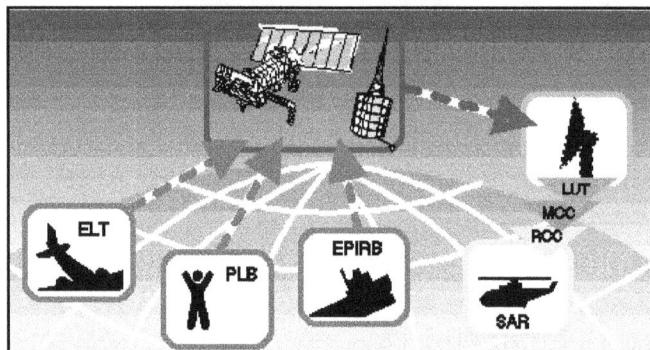

Fig 7.7 *ELTs, EPIRB, and PLB all transmit to a SARSAT-COSPASS equipped satellite. The signal is then relayed to a ground station which activates the appropriate SAR response.*

(7) Currently 406 MHZ systems are significantly more expensive
(8) A ground search should be started after 3 SARSAT fixes. Searches in likely areas may be started sooner if appropriate

3. NTAP statistics
  a. Retrospective analysis of known last NTAP with crashsite shows the aircraft typically found within 2nm of the last NTAP
  b. Experience indicates NTAP data is a powerful tool that justifies launching a ground search
  c. More detailed information of NTAP usage follows **(see page 7.9)**

4. Personal Locator Beacons (PLBs)
  a. Operate on 406MHz
  b. Must be registered so signal source is identifiable
  c. Can only be set off manually
  d. Currently only allowed by certain US Federal agencies and in Canada
  e. Pilot programs in Alaska, Oregon

Some downed aircraft crashes will ONLY be found by a ground effort.

VI.   Ground Operations

A. General sequence of ground deployment
   1. Missing aircraft report
      a. ELT hit- wait for additional satellite passes for refinement of plot. Check airports for false alarms. Consider mobilization of ground resources to a staging area if ELT related to overdue aircraft.
      b. General report of a missing airplane from citizen. Start investigation. Request NTAP data if appropriate.
      c. Citizen report of explosion- initiate air and ground search

> **Ground resources cannot be effectively deployed without a well defined search area, however mobilization and staging are logical first steps.**

   2. General area narrowed down to flight path
      a. Interview tasks on a county-wide area
      b. Wait for refinement of NTAP and SARSAT
      c. Check for ELT in air and if positive, on ground

   3. Ground deployment for searching requires defined area from ELT, NTAP, or sighting
      a. Establish ground Command Post/division in area.
      b. Start ground search techniques
      c. Helicopter operations justified

B. Direction Finding (DF) teams
   1. Obtain several bearings from roads and high places
   2. Obtain Bearings from multiple teams
   3. Only after lots of bearings send team into field
   4. Experienced operators and navigators critical
   5. Team in field requires excellent FTL

C. Investigation
   1. House knocking
   2. Maps on level of County Level. *Gazetteer* Statewide scale. Still appropriate to use ASRC/UTM grid overlay
   3. Communication problems and solutions
   4. Training in aircraft investigations ideal for team leaders
   5. Briefing to Team leaders should include the following points
      a. Introduce self, Uniforms and identification important
      b. Avoid leading questions
      c. Contact number for follow-up
      d. Do not give out too much information

ELT SIGNAL REPORT	1. Incident Name	2. Date Begun	3. AFRCC number	4. Page __ of ___

TIME	TEAM	LOCATION	BEARING (M or T)	DF TYPE	REMARKS

Fig 7.7 *Form used to track ELT fixes.*

     e. How to record information
        (1) Critical to record time of incident as reported by witness, Direction of flight if seen or heard. Anything unusual they noticed
        (2) Forms
     f. Thoroughness of task desired

D. Ground Tasks
    1. Special Team briefing
      a. Search techniques
        (1) Importance of looking up
        (2) Detecting odors and scent cones
        (3) Wider spacing
      b. Crash scene safety
        (1) Fumes
        (2) Fire hazards
        (3) Approach angles
        (4) Hot metal fragments
        (5) Infection control
        (6) Hazardous material
      c. Crash scene procedures
        (1) multi-casualty
        (2) crime scene
        (3) extrication
    2. Hasty -Ridges.
    3. Large sweeps- often based upon contours
    4. Wanderings
    5. Sweep and sector around last NTAP
    6. Expanding circle
    7. Dog usefulness
    8. Teams should be highly aware of odors. Attempt to run tasks to maximize wind

E. Medical and Evacuation Plan Considerations
    1. Possible need for extrication
    2. Number of high priority patients
    3. Possible remote location
    4. Likelihood of seriously injured patients
    5. Helicopters with winch capability and back-up plans
    6. Use of bulldozers

F. Scene Management
    1. IC must ensure all of the following are planned for before the aircraft is located
    2. On-scene coordinator
    3. Safety Officer
    4. Perimeter team (security)
    5. Evacuation team and leader

    6. Medical team and leader

    7. Psychological aspects

        a. Use of minors at crash scene

        b. Brief teams graphically on nature of scene

        c. Methods to reduce odors

        d. Limiting team's contact time

    8. Crash scene DOA management kit

VII. Clue Management

    A. Numbers usually quite high

    B. Most turn out to be erroneous

    C. Look for general patterns

    D. Investigation Form Handout

    E. Complex searches require aircraft-experienced ground IC

VIII. National Track Analysis Plan (NTAP)

    A. Introduction

        1. Software program used to look at backup tapes of RADAR data to determine the last plot on RADAR

        2. Original purpose as maintenance program.

        3. Requires an experienced supervisor (Quality Assurance Specialist) to operate and analyze the data. It is not an air traffic controller skill.

        4. How RADAR works

           a. Radio waves are transmitted from a rotating antennae, Antenna rotates once every 12 seconds

           b. Radio beam hits the aircraft

             (1) If the aircraft has the transponder turned on it will return the discrete code that is dialed into the transponder by the pilot. If the aircraft has mode C the aircraft's altitude will be sent back to the radar.

             (2) If the aircraft has no transponder a return signal will still be noted

             (3) NTAP data can be obtained from multiple radar sites. The computer program selects the radar site with the best data and only records a single set of data. The program does indicate when the radar site changes.

        5. Additional radar plots may come from military systems or some local approach control radar

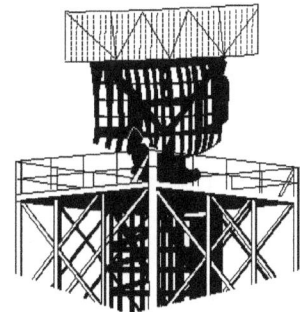

    B. Process Quality Assurance Specialist must use to determine last NTAP

        1. Four variations: Target time, target location, resolution, altitude

        2. Process greatly simplified when plane assigned a unique transponder number. Planes flying IFR are assigned unique numbers. Aircraft flying through controlled areas are assigned unique numbers

3. To determine plot must have a starting place and time
4. Once possible flight located, the last 20 or so plots are sent to AFRCC or the requesting agency
5. Printout of a single job request

C. The NTAP printout
   1. Three print outs of interest
      a. Figure legend- explains codes

      b. Map of flight- visual representation, overview of flight, has several built-in errors

National Track Analaysis Program Symbol Codes	
DATUM TYPE	CHARACTER PRINTED
LONG RUN LENGTH PRIMARY	+
SHORT RUN LENGTH PRIMARY	.
CORRELATED PRIMARY	X
UNCORRELATED BEACON	/
CORRELATED BEACON	\
MORE THAN ONE SHORT RUN LENGTH PRIMARY	=
MORE THAN ONE LONG RUN LENGTH PRIMARY	:
COMBINATION OF + AND .	&
MCI CORRELATED TRACK ELIBIBLE FOR CONFLICT ALERT	I
CORRELATED/UNCORRELATED VFR BEACON 1200 CODE	V
HIGH WEATHER SYMBOL	H
LOW WEATHER SYMBOL	L
SPECIFIC LOCATION OF REQUEST FIX	@
FDB, LDB, OR COMBINATION OF ABOVE	A-G, J,K,N-U
OVERWRITE - NO UNIQUE POLOT CHARACTERS AVAILABLE	?

FIG 7.8 *Symbols used in NTAP printout*

c. List three- gives the actual plots

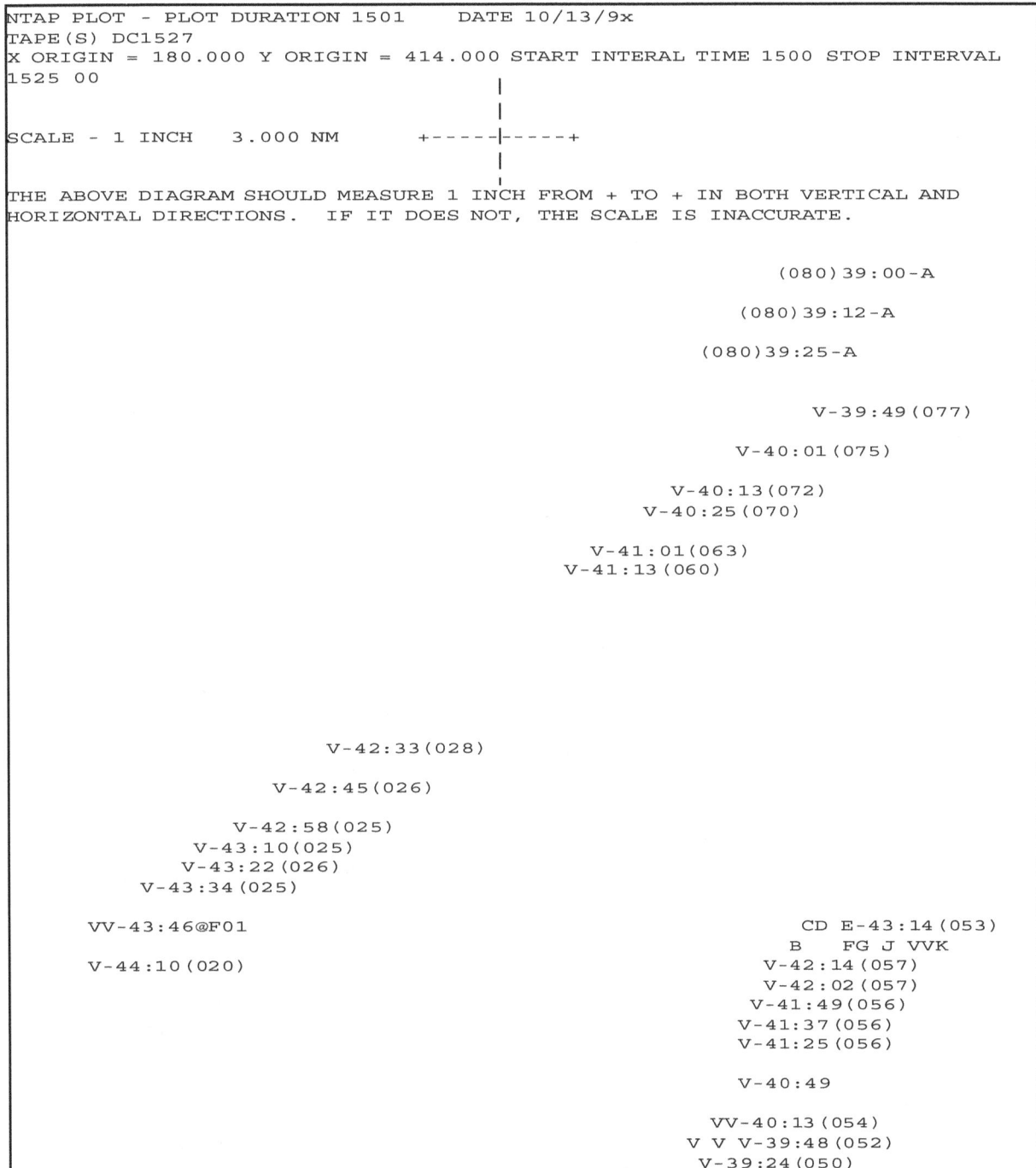

```
NTAP PLOT - PLOT DURATION 1501     DATE 10/13/9x
TAPE(S) DC1527
X ORIGIN = 180.000 Y ORIGIN = 414.000 START INTERAL TIME 1500 STOP INTERVAL
1525 00
                                        |
                                        |
                                        |
SCALE - 1 INCH    3.000 NM     +-----|-----+
                                        |
                                        |
THE ABOVE DIAGRAM SHOULD MEASURE 1 INCH FROM + TO + IN BOTH VERTICAL AND
HORIZONTAL DIRECTIONS.   IF IT DOES NOT, THE SCALE IS INACCURATE.


                                           (080)39:00-A

                                            (080)39:12-A

                                          (080)39:25-A


                                                V-39:49(077)

                                           V-40:01(075)

                                        V-40:13(072)
                                        V-40:25(070)

                                      V-41:01(063)
                                      V-41:13(060)




                    V-42:33(028)

                V-42:45(026)

              V-42:58(025)
            V-43:10(025)
            V-43:22(026)
          V-43:34(025)

     VV-43:46@F01                               CD E-43:14(053)
                                                 B   FG J VVK
     V-44:10(020)                               V-42:14(057)
                                                V-42:02(057)
                                                V-41:49(056)
                                                V-41:37(056)
                                                V-41:25(056)

                                                V-40:49

                                              VV-40:13(054)
                                            V V V-39:48(052)
                                                V-39:24(050)
```

Fig 7.10 *Visual representation of NTAP plot. List is used in conjunction with list three which supplies the detailed plots.*

2. Planning Factors
   a. 0.25 mile error of the radar
   b. Error increases when shifting from one radar site to another
   c. Radar capable of tracking cars, birds, etc. Each radar site configuration is different. Only one sweep required to see aircraft
   d. Ghost plots

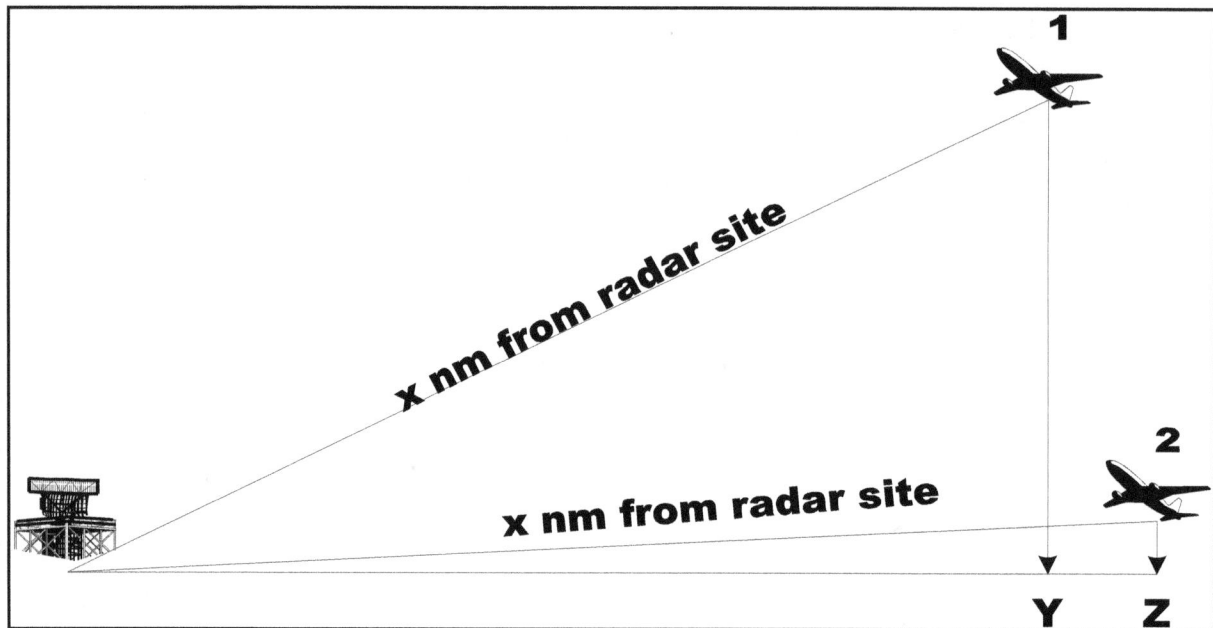

Fig 7.9 *If the plane's transponder is incapable of reporting altitude the radar site will only be able to calculate the distance between the radar site and the target. The greater the altitude the greater the error. Plane 1 and 2 are the same distance from the radar site. Therefore, they both will be reported to be at position Z. The error is the distance between point Y and Z. Fortunately for SAR planners the final NTAP plot is usually at low altitude. For planes that breakup in mid-air, searchers will face additional difficulty.*

    e. Missing plots

    f. Radar are aligned and calibrated daily

    g. Altitude-encoded plots are more accurate because computer calculates a ground range. Without altitude encoding (Mode C) slant range is calculated. The greater the plane's altitude and the greater the distance from the radar sight, the greater the error.

3. Ground Planning process

    a. Plot the last 15 plots (3 minutes) onto a 7.5-minute topographic map

    b. Calculate airspeed

    c. Calculate vertical change

    d. Seek input from expert if any of these speeds or changes in altitude exceed the structural parameters of the aircraft.

    e. If unsure of the ceiling of the radar or shadows, consider sending an aircraft to recreate the flight path

    f. Examine the map plot carefully for possibility of discontinuous track

---

To Calculate Airspeed

* Measure the distance between two NTAP plots in miles.
* Multiple the distance by 300 to obtain miles/hour (mph).
* Multiple mph by 0.87 to obtain nautical miles/hour (knots/hour)

* To directly obtain knots/hour from mile measurements you may multiply by 260.7

Formula based upon a 12 second difference between the last two plots. In some cases a plot may be missing and 24 seconds lapse between plots.

UNIT 8: INCIDENT COMMANDER ROLES AND RESPONSIBILITIES

Objectives:
- ☐ Describe the roles and responsibilities an IC may perform; upon being assigned as IC, enroute to the incident, upon initial arrival, after initial tasks, during the second operational period, and at the end of a search.
- ☐ Describe important considerations in selecting a replacement IC and the process of new IC transition.
- ☐ Describe the role of the Incident Commander in relation to the Legal Responsible Agent/Agency Administrator (RA/AA) when the RA is uncooperative and when the mission involves or expands into other jurisdictions.
- ☐ Describe the role of the IC in relation to the various resources that may participate in a search mission when the IC has overall responsibility for all resources present or when there are resources present that may not be willing to cooperate.
- ☐ Demonstrate the ability to communicate with staff by means of briefings, meetings, and written communications.
- ☐ Describe the internal staff information flow system, including verbal, written, and electronic communications, which are required to insure that information is properly collected, evaluated, disseminated, utilized, and stored throughout the incident.
- ☐ Describe common mistakes inexperienced Incident Commanders make and how to avoid them.
- ☐ Identify outside influence problems common to search missions such as family, media, politicians, and psychics. Describe solutions and justify the solutions.
- ☐ Describe the process used in making the decision to suspend a mission.
- ☐ Explain the IC's role after the subject or target has been located.

I. Introduction
A. Relationship between IC and staff
B. Nothing truly prepares for the role
C. Topics covered: operational duties, external influences, IC-RA interactions, common mistakes, and suspensions

> The IC is responsible for everything. Do not blame others for failures. Even circumstances beyond your control are your responsibility.

> The IC often receives the praise for searches that go well even if the IC had little to do with the success. Expect the reverse.

Initial Response Resources
- Management (Overhead Team)
- Tracking dogs
- Air-Scent dogs
- Trackers
- Hasty Teams
- Helicopters

II. Operational Duties

A. Upon being notified of Incident Commander role
  1. Obtain brief subject briefing
     a. Age, sex, physical and mental condition
     b. PLS, LKP, time last seen, circumstances of loss, weather, adequate clothing
  2. Obtain current and predicted weather information
  3. Obtain directions to the incident and check them personally against a detailed map
  4. Obtain a phone number and name of the Agency Administrator (AA) or Legal Responsible Agent (RA), the State Coordinating Officer (SCO), and the current command post
  5. Determine number and type of resources on scene, what has been done, and if the Point Last Seen (PLS) has been preserved
  6. Determine level of response needed by coordination center or AA
     a. Overhead team- 2-4 management-qualified members with IC
     b. Quick response team- overhead team and 5-10 Field Team Leaders (FTL)
     c. Full callout- unlimited to specific target size
     d. If any further modification to these levels is required
  7. Request appropriate Search and Rescue trained resources (Field Operational Units, SRU, or FOU) be placed on alert or asked to mount a simultaneous response
  8. Determine if air travel required and place request with an appropriate agency
  9. Request from the dispatcher that appropriate equipment be dispatched to the incident (copier, maps, radios, rescue gear, etc.)
  10. Tell dispatcher what your personnel requests are at this time, and your projected request upon reaching base
  11. Tell the dispatcher your estimated time of departure, when you actually leave, and ETA to search base

N.B. Depending upon the circumstances, talking directly to the coordination center and/or AA may be necessary

B. While enroute to the search
   1. If communication unavailable check in with dispatch once an hour
   2. If two IC qualified individuals are traveling together determine who will serve as IC before departing and then inform dispatch of decision
   3. If cellular or radio communication is available, receive updated information from the coordination center and dispatch before arrival at incident, and inform when IC arrives on scene
   4. Determine resources enroute, place appropriate groups on alert, start arranging logistical support. Update directions upon arrival.

Fig 8.1 *Cell phones should be used to increase the working time of responding ICs.*

C. Upon arrival

The precise order of tasks upon arrival will change depending on the unique circumstances of the search. The IC must be highly flexible during the first hours of an incident and the following list serves merely as a suggestion for initial tasks. The precise order of theses steps will vary.

   1. Meet AA or current IC and introduce self and other overhead staff members
   2. Receive initial overview briefing from AA or incident briefing from prior IC (Incident briefing ICS 201). Review documentation of tasks completed or in progress
   3. Appoint investigations and operations chief (OPS) from most qualified individuals present and allow them to begin their jobs. Activate other elements of the Incident Command System (ICS) as needed
   4. Delegate sign-in procedure to competent individual
   5. Continue with more detailed briefing from AA covering
       a. How has the PLS been protected?
       b. Media policy
       c. Expectations of IC and AA
       d. What AA wants done right away and later?
       e. AA analysis of foul play, suicide, violence possibilities
       f. How SAR resources operate
       g. Limitations of AA and IC
       h. Need for written delegation of authority
       i. Use of minors
       j. Establishing communication channels (i.e., keeping each other informed)
       k. Possible search scenarios based upon similar searches
       l. Status/needs of family members
       m. How to notify family if subject found DOA?
       n. Status of investigation function

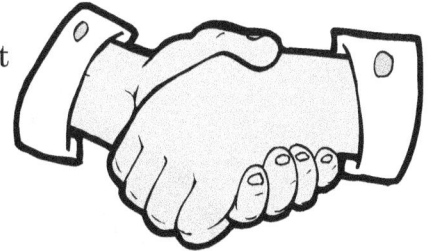

    o. Types and numbers of SAR resources
    p. Recommendations of what SAR resources to call
    q. Medical coverage in area
    r. Potential for discovery of crime scene?
    s. Procedure if crime scene discovered
    t. Any hazards in area?
    u. Security of area and any risk to searchers?
    v. Any other special procedures and requirements (policy, jurisdiction, wilderness area limitations, etc.)?

6. Brief command and general staff

> **Inadequate bases almost always ensure inadequate searches.**

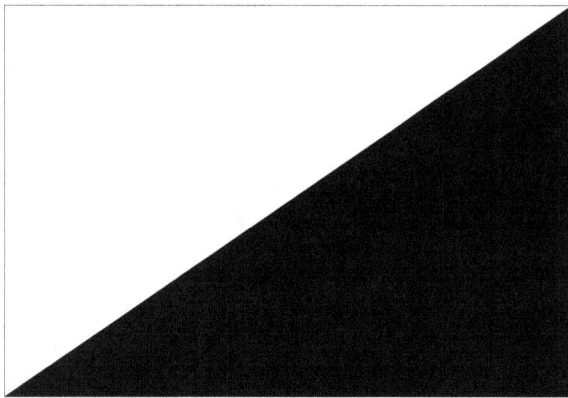

7. Discuss with AA moving base if minimum base requirements are not met at current location
    a. Electricity, lights (generator may provide)
    b. Work area sheltered from weather, media, family
    c. Telephone (if not brought)
    d. Copier (if not brought)
    e. Radio communications (if not established)
    f. Running water if large or multi-operational period search
    g. Sanitation if large or multi-operational period search
    h. Staging area if large search
    i. Parking area if large search

8. Determine information needs and inform applicable personnel of needs
9. Develop initial objectives /strategies (develop or delegate tactics)
10. Delegate task of making and copying subject information sheet. Approve draft before copying and distribution. If possible have AA review and approve draft
11. Coordinate staff activity
    a. Manage incident
    b. If appropriate, ensure that operations is developing tasks and dispatching teams within 30 minutes of arrival
    c. Approve the use of different training levels on the incident.
    d. Ensure efficient flow of personnel from staging area to field
    e. Constantly be available to general staff
    f. Work to maintain high morale and confidence
    g. Work to create an environment for staff to work in, i.e. get things

done. Shield staff from family, media, and political pressures
- h. Drive mission, identify problem areas, influence overall direction of mission, establish initial objectives, and enforce priorities
- i. Verify that everyone on general staff is being kept up-to-date
- j. Establish parameters for general staff members to increase staffing levels

12. Coordinate and determine resource requirements
    - a. Determine OPS needs for any resources
    - b. Working with AA, place any additional local resource requests
    - c. Working with other SAR resources on scene, place additional resource requests of those groups. Ensure the safety and health of SAR resources already deployed
    - d. Keep AA and coordination center informed of resource requests

13. Coordinate and manage external factors
    - a. Contact Coordination center, brief on situation, update directions, place resource request, inform of resources on location, inform who (by name and position) will be placing resource request in the future (IC, OPS, RESTAT, etc.)
    - b. Contact dispatch, brief on situation, update directions, place resource request, place equipment request, obtain incoming equipment and personnel lists. Inform if any restrictions on minors, training levels, or numbers.
    - c. Identify and meet with liaisons/agency representatives to keep informed and determine their special capabilities and requirements.
    - d. Ensure tentative medical/evacuation plan developed in conjunction with local EMS/Rescue

D. After initial tasks
   1. Keep AA, SCO, Coordination center informed of all important actions. Keep AA, SCO updated at least every two hours when present or at an agreed-upon schedule
   2. Ensure staffing levels meet needs of incident
   3. Ensure staff meeting paperwork and documentation requirements
   4. Ensure staff not top heavy, or reduce staff after the initial tasks are dispatched
   5. Evaluate logistical needs, future needs, and ensure they will be met
   6. Determine when the second operational period will occur
   7. Determine need for formal Incident Action Plan (IAP) including at least
      - a. Search objectives
      - b. Personnel assignments
      - c. Medical/evacuation plan
      - d. Radio plan
      - e. Demobilization plan
      - f. Subject information sheet
   8. Management by wandering

E. Second operational period
1. Approve and authorize implementation of IAP
2. Ensure planning function met
3. Ensure logistical needs met
4. Ensure resource orders placed
5. Ensure family and other liaisons maintained or expanded
6. Ensure staff members rested enough to be safe and functional
7. Ensure flow of incoming personnel to field deployment functioning smoothly
8. Determine when, where, and with whom to hold planning meeting

F. **Changing IC:**
1. When to Change Incident Commanders
   a. On type one and two NPS teams the IC stays in place for three weeks

> **Most deaths occur on fire incidents when a type III IC delays the transition to a type II IC.**

   b. During the first 48 hours 3 hours of sleep/night normal. Afterwards, little to no excuse to get less than 6. IC remains in command while resting. Deputy IC or operations typically "in charge" while absent.
   c. Reasons for an IC to step down
      (1) Search complexity is increasing. IC feels a higher level IC is required
      (2) IC must leave due to work or personal commitment
      (3) A second incident occurs
      (4) IC feels another IC is better able to contribute to search management
      (5) IC feels they have lost their effectiveness with staff, RA/AA, or external influences
      (6) IC feels they have lost interest in the search and can no longer contribute
      (7) IC removed by request of RA/AA, state agency, or own agency
   d. Deciding upon the replacement IC
      (1) Decision of planning staff with ultimate approval of the current IC or AA
      (2) Factors in choosing the replacement Incident Commander include
         (a) Type & experience of replacement IC
         (b) Availabilities of replacement ICs
         (c) Current rest level of replacement IC
         (d) Length of time the replacement can commit to the search
         (e) Current position or familiarity with the search replacement IC currently has
         (f) Past experiences the replacement IC has with working with RA/AA
         (g) Meeting recertification requirement

2. Process to change Incident Commanders
   a. Current IC or AA selects new IC from IC-qualified personnel
   b. If possible IC designee should be present and unassigned for an operational period before transition
   c. Current IC must brief IC designee
   d. IC should introduce IC designee to AA, SCO, other important agency representatives, and staff
   e. IC should inform AA, SCO, agency representatives, coordination center, and dispatch what time IC designee taking over
   f. The exact moment the IC designee becomes the new IC must be clear to both ICs and the general staff. The exact moment of transition should be announced on the communication network and entered into the communication log, operations log, and IC unit log.

G. End of search procedures
   1. Report finds or suspensions directly to AA, dispatch, and the coordination center. State what actions should be taken
   2. Ensure other SAR resource's dispatch informed of find or suspension
   3. Activate and ensure demobilization plans operating smoothly
   4. Arrange, if possible, on site end-of-mission debriefing among staff members to review good points and learning points of mission
   5. Arrange, if possible, on site end-of-mission debriefing to all personnel
      a. Cover find information
      b. Subject condition
      c. Operation overview
      d. Credit all organizations
      e. Provide stress debriefing information
      f. Review demobilization procedures
      g. Final safety briefing

   6. Complete or delegate the task of filling in incident summary report
   7. Provide AA, SCO with copy of key paperwork
   8. Determine if any SAR resources still enroute to base
   9. Ensure personnel well rested and fed before releasing
   10. Inform coordination center and dispatch when base closes
   11. Ensure copy of paperwork sent to archives

III. Special circumstances

A. Arrival to high urgency search with minimal overhead team
   1. Collect initial information from AA
   2. Start sign-in procedures
   3. Have AA organize all undeployed resources together and give subject information briefing
   4. Have SAR overhead team member give briefing on SAR procedures (normal FTL briefing)

5. Evaluate search for safety and clue potential
6. Select "FTL" from established leadership of available resources
7. Rapidly develop tasks using larger number of team members than normal. Attempt to assign medical personnel to each team
8. Brief "FTL" on task and FTL procedures
9. Dispatch teams
10. Assume OPS function or delegate to SAR-trained personnel
11. Delegate investigation function to law enforcement
12. Utilize existing communication infrastructure

B. Unable to find replacement IC
   1. Determine if on-scene resources or incoming resources lacks an IC-qualified person
   2. Inform dispatch of IC deficiency and request that all ICs be contacted at least twice and informed of acute need
   3. If deficiency still exists inform coordination center of problem
   4. If deficiency still exists have dispatch contact an Area Command Authority (ACA) qualified person. The ACA may then appoint an agency representative, an interim IC, transfer responsibility, or suspend the mission.

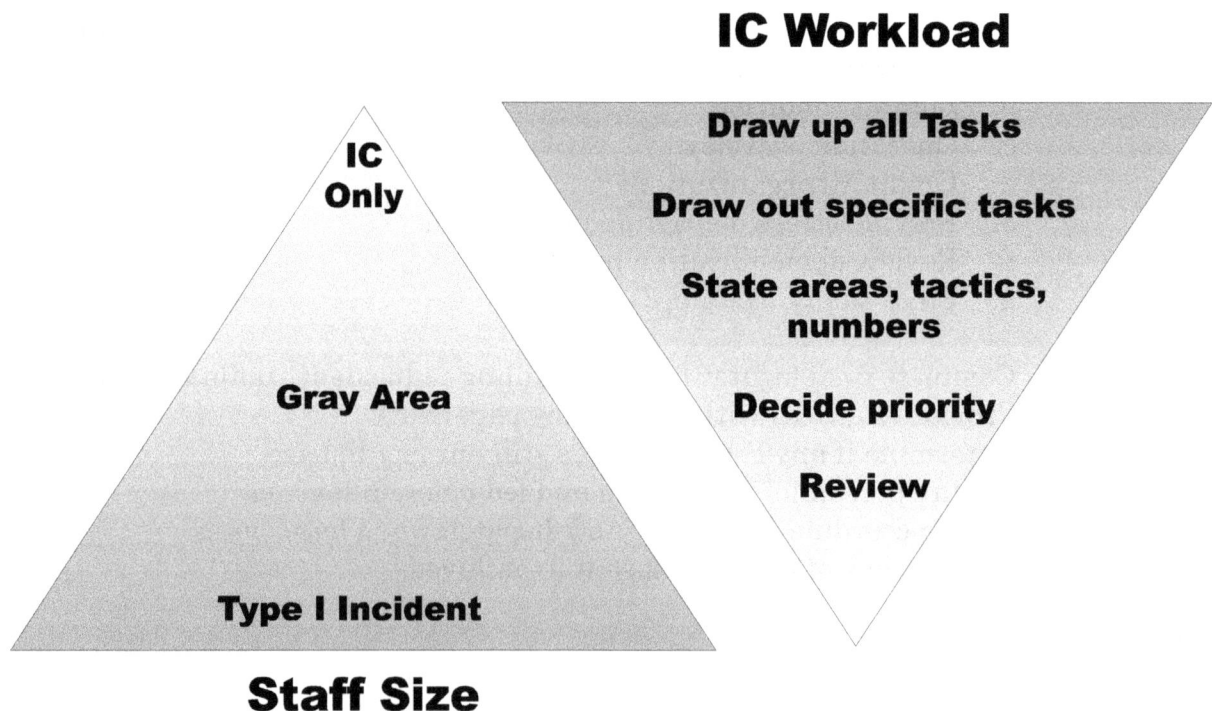

IC Workload

IC
Only

Draw up all Tasks

Draw out specific tasks

State areas, tactics, numbers

Gray Area

Decide priority

Review

Type I Incident

Staff Size

C. Amount of tactical guidance to staff
   1. Tends to be greater in beginning of search
   2. IC generally most experienced person on search
   3. Need to set staff in motion
   4. Depends upon experience of staff
   5. Fun to keep hand in tactics

    6. Important to trust staff
    7. Levels of staff guidance
       a. Draw out actual initial tasks
       b. State high priority areas, type of tasks, and number of tasks
       c. Decide initial priority of staff-drawn tasks
       d. Review staff-drawn initial tasks
    8. Review why staff slow

IV. External Influences

A. Core Philosophy
    1. Prevent problems (proactive)
    2. Communications=Tell your story> Dig for problems
    3. See problems developing by developing information network
    4. Find hidden agendas of all players
    5. Find the time to talk in a causal manner to many key players

B. Other Trained Incident Commanders
    1. Cause stress
    2. Good source of advice and idea bouncing
    3. Often bring fresh outlook based upon their previous experiences
    4. Best role difficult to decide
       a. In base
       b. In field
    5. Require a separate briefing

C. AFRCC
    1. Recent trend of direct contact (nature of contact)
    2. Background information on AFRCC
       a. Federal regulations
       b. Controller training
    3. SAR Operations
       a. State SAR Agreements
       b. Posse Comitatus Act
       c. Medical Evaluation
       d. Conflict of Interest
       e. Resource availability
       f. Urgency of situation
    4. Some recent politics
    5. Mission not closed until all "field resources returned"
       a. Normal mission closing procedure
       b. Importance in air transport situations

D. State SAR Coordination Agencies
1. SAR Duty Officers
   a. Resource orders, updates, problems, finds
   b. Legal constraints often limit options
   c. Need to push air-transport resource orders
   d. Be prepared for "devil's advocate" role
   e. Seek advice, bounce ideas, paint picture, be honest

2. Others
   a. Watch officers
      (1) Answer the phone at Emergency Operations Center (EOC)
      (2) SAR training received
      (3) EOC role
      (4) Dependability

   b. Regional Coordinators
      (1) Locations
      (2) SAR training received
      (3) Logistical connections

   c. Command Bus personnel
      (1) Logistical capability
      (2) Required lead time
      (3) Training of personnel

E. Agency Representatives
1. Identify
2. Give a personal greeting and briefing
   a. Give your version of the story
   b. Acknowledge their importance
3. Hurry along
4. Determine any special needs
5. Possible inclusion in planning, suspension, etc.

F. Family/Friends
1. Role of Liaison officer/Importance of keeping family happy
   a. Intense emotional state
   b. Importance of getting SAR story out
   c. Importance of information and uncovering hidden information
   d. Minimize negative press, lawsuits
2. IC distancing from family
3. Beware of factions
4. Be honest: give hope with judicious omissions
5. Protect family from the emotional roller coaster of searches

6. Protect staff from family
   a. Why family should not be in base for protracted periods
      (1) Stress on staff
         (a) need for black humor
         (b) need to create atmosphere staff can freely exchange ideas, criticisms
      (2) Stress on IC
         (a) same as above
         (b) best not to get too close to family for suspension decision.
      (3) Stress on family
         (a) avoid roller coaster
         (b) do not understand what is going on, what is said
         (c) viewing people in base
            i) lots of people in base: "why aren't they in the field looking"
            ii) few people in base: "why doesn't anybody care, where are they all"
7. Family/friends in the field
   a. Initial
      (1) important to keep family around base for investigation and verification of clues
      (2) family members on teams cause considerable stress to field team leaders and members, who often are less trained to handle the additional stress
      (3) Desire to keep family member from possibly finding dead or injured subject
      (4) Family members often ill equipped, not trained to go into the field
   b. As search progresses
      (1) Investigative function/explanation decreases
      (2) Family better mentally prepared to find deceased subject
      (3) Stress of sitting around base and not doing anything increases
      (4) Family/friend who are in good physical shape often obtain or purchase necessary equipment to go out into the field
      (5) Many family members should never be allowed into the field due to mental and physical health, lack of equipment, or dangerous terrain that requires training
   c. Advantages of Family/friend in field
      (1) They are out of base
      (2) Increased understanding of what searchers are facing
      (3) Satisfaction of contributing to the search
      (4) Physical activity helps mental stress
      (5) Often will search in the field with/without permission of search management. Increase overall safety by sending with "baby sitting" team leader

G. Media
   1.   Importance of being proactive
   2.   Remind everyone of press policy
   3.   Roles and responsibilities of Information Officer covered later in course
   4.   Need to change attitudes towards press. Media is a resource tool for you to use (fire-type resource)

   5.   Choosing press "color interviews."
      a.   Mature, responsible, does not shoot from the hip
      b.   Hold test interview, complete with traps
      c.   Clearly establish parameters with press and interviewed; brief on press traps
      d.   Dog handlers often have considerable experience with color interviews
      e.   Field tasks

H.   Psychics
   1.   Review MSO/MSF/MLPI textbook
   2.   Log as clues, how much weight given depends
   3.   Often a day three phenomena

I.   Politicians

   1.   Another type of fire resource
   2.   Need to be kept well informed
   3.   Well organized-briefings including: objectives, strategies, resource expectations, expected problems and needs (they can often solve some of these problems and take satisfaction from helping).
   4.   Best to give some type of resource request
   5.   Candid honesty best policy; can make powerful ally if you make them look good
   6.   Remember, to a large part, the role of the IC is political

V. IC-RA/AA interactions

A. Initial meeting
  1. Information collection technique
      a. Opening line
          (1) I'm the person the state sent
          (2) Desire to help
          (3) What can I do for you
          (4) Often avoid Incident Commander title at first
      b. Allow RA to give available information.
      c. Ask questions to demonstrate you are listening and knowledgeable
      d. Continuing asking questions to a sophistication level to show RA you have ideas of what needs to be done. STOP! Once you have made this point. No need to make RA/AA look bad or ignorant

> **Professional relationships take time, trust, work, and often a sense of humor to develop.**

  2. Reassurance and professionalism technique
      a. Draw parallels with past searches
      b. State press policies to demonstrate volunteer resources are professional and do not represent loose cannons
      c. Share information
      d. Prevent miscommunication

> **Buzz words an IC wants to hear from the RA/AA:**
> **"Its great to see you again, help!"**
> **"You're the expert, do what you think is best."**

  3. Demonstrate action, ability to make decisions
      a. Have staff members in wings, make appointments, give tasks with short sentences
          (1) operations officer
          (2) investigations
      b. Discuss resource order recommendation

B. Follow-up communication
  1. Consider approval of Subject Information Sheet (SIS).
  2. Give updates on resource requests.
  3. Give update on field deployments.

VI.   Staff Management Techniques

> **With proper guidance and direction the staff should do all the work for you.**

A. General Tricks
  1. Management by wandering.
  2. Initial IC role intense during first hour. Make sure key players are identified. Make sure staff knows players who can help them.
  3. Stay visible.
  4. Search size changes ratio of managing versus running.
  5. Monitor data flow
     a. Past problems
        (1) Data transfer failure points
        (2) Field resources or lower staff member make judgements
     b. Prevent problems
        (1) Monitor communications, look over communications log to verify all clues followed-up and entered into clue log
        (2) Listen to debriefings; make sure any clues stated recorded on debriefing sheet and clue log. Make sure recommendations recorded on follow-up log
        (3) Question clue unit leader or operations, can they state knowledge of the action taken on all reported clues
        (4) Question investigations. Make sure all reported sightings and phoned in clues are recorded and followed up
        (5) Train staff to develop proactive search for problems also
  6. Primary function is to develop staff structures
  7. Amount of supervision difficult to decide
  8. Need to pressure staff
     a. Why staff slow
     b. How to speed staff up

B. Planning Meetings
  1. Advantages
     a. Allows planning
     b. Allows coordination and sharing of information
     c. Allow to blow off steam
     d. Gives a break from the search
  2. Disadvantages:
     a. Take time away from the search
     b. RA/Family/Field resources often do not understand
  3. Points
     a. Keep short
     b. Make sure everyone is prepared
     c. Make sure no negative impact on running search
     d. Release members when possible
     e. Good to invite, RA, key experienced personnel

VII. Common Mistakes of New Incident Commanders

A. Obsession with politics
   1. May have to disagree with RA (common)
      a. RA wants to close search too quickly
      b. May not want to escalate search
   2. Afraid to make decisions and stick with them
      a. OK to change mind based upon new information but avoid constant flip-flopping

B. Lose sight of IC role
   1. Maintain more comfortable operation role
   2. Jump into operational tasks
   3. Get tied down
   4. Purpose of not being tied down
   5. Failure to give staff guidance and instruction

> **Personal Goal: I want to be able to tell the family I did everything reasonably possible, and mean it.**

C. Failure to make search grow
   1. Tendency to work with what you have
   2. Review the ideal planning process
   3. Strive to be proactive vs reactive
   4. Failure to place timely resource orders
   5. Failure to involve local resources if required

D. Lack of inspirational vision
   1. Drive mission, get teams out
   2. Proactive thought process
   3. Need clear idea of both what you want to accomplish and transmit that view to staff
   4. Need to develop methods to measure effectiveness during the search
   5. Lack of following operational philosophy

E. Loss of control over data flow
   1. Documentation not being completed
   2. Clue/information loss
   3. Failure to seek/push investigation

F. Fatigue
   1. One of the most common mistakes
   2. Failure to trust staff
   3. Common mistake even among experienced staff and IC
   4. Need to force staff members to sleep and recognize own limitations

G. Improper staff selection
    1. New Incident Commanders often work with more untrained staff
    2. Type III Incident Commanders may not even have staff
    3. Failure to match staff personality with appropriate job
    4. Avoid staff selection based upon politics

H.   Overreaction to clues/information
    1. Locking into one scenario
    2. Overreaction and switching of resources
    3. Obsession with clues
        a. Most clues have no relationship to subject
        b. Most clues identified are not critical
        c. False positives
        d. False negatives

No mission is a training mission; yet all missions involve training.

I. Failure to switch Incident Commanders

VIII.   Finds/Rescues: the Incident Commander Role

A. Background
    1. Squirrel; charged; suppressed ego appear; need for action, switch to rescue mode. Several other tasks also required
    2. Need for fast decisive leadership. But must take time to think. A rash decision that allows a poorly led or equipped evac/medical team to depart into the field. Failure may lead to hours of delay.

B. Core elements
    1. Activation of Evac/medical plan
        a. Leader selection
        b. Team composition selection
        c. Equipment selection
    2. Control of non-SAR resources
    3. Notifications
        a. Subject confirmation issues
        b. State agency, Dispatches, Family, Responsible Agent, Press

IX.   Suspensions

A. Suspension overview
    1. AA often pushes for suspension prematurely
    2. Current SAR teams tend to suspend earlier than past searches
    3. Most difficult decision an IC can make. Will often haunt the IC years later

B. Suspension Meeting Process
1. Attendance
   a. IC, Command staff, General Staff, RA, key agency representatives
   b. Must not include family or friends

2. Time. Generally takes 30 minutes to one hour. Make sure the incident is taken care of

3. General Process
   a. Review the following information
      (1) How search area defined
      (2) Assessment of search area coverage
         (a) PODcum on all sectors/shifting POA
         (b) Sectors researched
         (c) Mixture of resources
      (3) Clues or lack thereof
      (4) Investigation results
      (5) Subject survivability/weather
      (6) Safety of searchers/weather/fatigue/equipment
      (7) Costs
      (8) Available resources/other missions
      (9) Family pressure
      (10) Political Pressure
      (11) Gut instinct
   b. Use the factor list described in MSO/MSF/MLPI text
   c. Review results of factor list. Useful for someone to play devil's advocate
   d. Come to a consensus on suspension decision
   e. Document process well. Best to take notes. Tape recording is not recommended

4. Practical factors to consider
   a. Day three phenomenon
   b. Survivability
   c. Evasive subjects
   d. Limited resources

C. Decision Shifting
1. The big push "weekend concept". End Sunday 2-4 PM. Allow sufficient travel time for regional SAR resources to return home safely
2. Shifting final IC
   a. IC #1 will suspend if "nothing significant is found"
      (1) #2 actually makes the decision to suspend
      (2) Actual suspension is not performed, allows a mental "out"
   b. IC #2 Takes over
      (1) Tends to feel suspension decision has already been made so it is not "their" decision. They simply are carrying out an existing plan

(2) Still feel empowered since able to reverse decision

(3) Tends to be far less stressful

(4) Second IC often brings fresh perspective and closing enthusiasm. Knows he/she will be viewed as a hero if an unexpected find is made

D. Press observations
1. Be honest
2. Press has never been surprised or negative about suspension. Tends to question search if considerable funds are being spent
3. Comments to press; we search as long as.....

E. Limited continuous search
1. Emphasis the change in focus of the search from field resources to investigation. The search is continuing
2. SAR field resources will remain on call in case of further developments or clues

F. Methods to prepare the family
1. 2-3 days before possible suspension, drop hints that search cannot go on forever
2. Show family that search objectives are being met
3. Drop hints that resources are finite
4. Big push concept
5. Gradual removal of resources instead of all resources leaving at once
6. Search shifting from field to investigation

UNIT 9: COMMAND STAFF

I. Information Officer (IO)

The Information Officer (IO) is responsible for the formulation and release of information about the incident to news media, other appropriate agencies and organizations, and to incident personnel.

- A. Activation by IC
    1. Whenever media contacts must be initiated by search effort
    2. Whenever significant media contacts begin interfering with incident
    3. Whenever IC appoints
    4. Position may be combined with other command position
    5. Position Qualification: IC, Incident Staff, specific IO training

- B. Roles and Responsibilities
    1. Obtain briefing and report to IC
    2. Work closely with AA/RA to develop media policies and plans
    3. Develop media policy including subject find procedure for inclusion in IAP
    4. Create a single incident information (media) center if possible
        a. Establish early
        b. Make sure clearly identified
        c. Site removed from incident scene but close enough to command center
        d. Large room for briefings
        e. Good lighting
        f. Telephone access
        g. Work area for reporters

Objectives:
- Describe the roles and responsibilities of the Information Officer, Safety Officer, Liaison Officer, and the Family Liaison Officer.
- Describe methods to give an effective media interview.
- Identify potential safety issues and describe how they can be mitigated if possible.
- Describe when risk factors outweigh the need to continue operations.
- Describe the common signs of incident stress and define the criteria for recommending a critical incident stress debriefing.
- Describe methods to keep the family informed.

A professional should not tremble at the sight of a camera but instead should easily control the media.

Problem:
List reasons to welcome media attention.

Media Plan	¹ Incident name Shuping	² Date prepared Dec 17, 1997	³ Time prepared 11:00	⁴ Operational Period 06:00-20:00
⁵ Media Contacts ☐ additional sheets				
⁶ Media Organizations	⁷ Contact person		⁸ Contact number	⁹ Deadlines
WMAL (Radio)	Sue Wave		555-3659	11:30, 16:30 22:30
Daily Info (Newspaper)	Winnie Pennington		555-2422	18:00
WTVN (TV)	John Micheal		555-1324	11:15, 16:15, 22:15

¹⁰ Regular Briefings			
Times	Participants	Location(s)	Preparations
As needed			

¹¹ Media Procedures for Subject(s) found alive and well
1) Confirm correct subject, 2)Contact family first, Family has given permission for a happy reunion photo op. 3) Prepare statement for press. 4) Facillitate reunion photo op. 5)Schedule summary press breifing. Sheriff wants to be present.

¹² Media Procedures for Subject(s) found injured
1) Confirm correct subject, 2) Contact family first, family asks to be protected from media in case of injury 3) Prepare statement for press 4) Prepare roadhead photo opportunity press area try to keep media from going into the field. 5) Schedule summary briefing, Sheriff wants to be present.

¹³ Media Procedures for Subject(s) found deceased
Same as above

¹⁴ Additional Procedures

Media plan dh8 302 1/96	¹⁵ Prepared by (IO) Julie Talkwell

Fig 10.1 *Media Plan*

> **Unprofessional buffoons should fear the media the most, but often love to hear and see themselves.**

Fig 10.2 *Professional dress and behavior are strongly recommended for the IO and IC*

h. Bulletin board, chalk board, or flip chart
i. Sufficient electrical supply
j. Refreshments
k. Restrooms

5. Working with Logistics (LSC) arrange needed supplies
6. Obtain additional briefing from OPS, PSC, documentation, investigations, and RESTAT
7. Determine any constraints on release of information from IC and AA
8. Obtain approval of information to be released from IC. Work with investigator to determine if some key subject information should be kept from media for later investigative verification
9. Arrange media conferences at announced times as needed
10. Release news to media and post information in command post and other appropriate locations
11. Develop a list of media contacts (TV, radio, print)
12. Make initial contact with media if IC requests
13. With IC approval, arrange meetings between media and incident personnel
14. Help develop a biographical sketch of IC
15. Help develop or distribute fact sheets on participating SAR organizations
16. If approved, provide escort services for media into the field and command post
17. Respond to special requests for information
18. Attend staff meetings to stay informed
19. Monitor news media releases. Videotape, record, or clip papers if possible
20. Keep IC informed of any adverse reactions
21. Help prepare IC for any interviews
22. Investigate any rumors and attempt to obtain facts
23. Keep searchers informed of developments in search
   a. Post map showing area covered

       b. Post Subject Information Sheet (SIS)

       c. Post search objectives

       d. Post important information (lodging, food, news clippings)

24. Ensure patient status not released before next of kin contacted
25. Help shelter family from media if they request

C. Interview Techniques
   1. Developing your interview answers
      a. Important to develop your own message that you want to get out. Similar to establishing objectives and then strategies (message). Objectives should be positive. Messages should
        (1) Be brief and use simple, plain language
        (2) Have the message up front, emphasizes benefits
        (3) Be story-like, make people oriented, personal experiences
        (4) Use few negative words
        (5) Don't be disparaging towards any organization or individual

Event: During a semi-technical evacuation a member of the rope team, not wearing their helmet, falls and is knocked unconscious. This requires a second evacuation which delays the initial evacuation by two hours.

Main Points (Objective)	Story/Example/Description

   2. Don't try to memorize response

   3. Give 10-20 second response

   4. Single message in each response

> **If you can't be short and to the point, you don't know your subject well enough to communicate it!**

   5. Controlling the interview
      a. Packaging/Bundling
        (1) Quantify your information and tie it together giving the reporter verbal clue to follow
        (2) Example. "We have three different search scenarios we are using for planning"
      b. Bridging
        (1) Verbal maneuvering that allows you to "dodge" a reporter's question and move on to the message you want to get across
        (2) Bridge does not have to be fancy but it must be valid

> **Bridging Examples:**
> **"What concerns me even more..."**
> **"In my experience..."**
> **"The critical issue is..."**
> **"That's one perspective..."**
> **"I've heard that, but the real focus should be..."**

6. Interview Do's and Don't:
   a. Deal with the media like you would want others to treat you
   b. Media want information, whether you cooperate or not, they will do a story. Be aggressive with the media
   c. Give the main objectives and then support them. Don't feel you have to keep talking. Make the interviewer keep the conversation going.
   d. Don't answer with just a simple "yes" or "no" or "no comment." Don't be curt, there is no such thing as a dumb question
   e. The best defense is a good offense. Be proactive. Let your deeds speak. Accentuate the positive
   f. Discuss only matters you have direct knowledge of. Avoid hypothetical situations. Remember, there is no such thing as a personal opinion when you are speaking for all SAR resources
   g. You aren't obligated to tell everything you know. Some things are better left unsaid
   h. If you can't answer the question, give a reason why. There's nothing wrong with "I don't know." Avoid "No comment"
   i. Don't pretend to be perfect. Admitting mistakes from time to time demonstrates candor and the integrity of your organization.
   j. Complete your answers within 30 seconds
   k. Do not use acronyms or technical terms
   l. If you wish to change or drop a topic, bridge to another topic
   m. Take time to think about the question. If you are not clear on the meaning, ask to have the question repeated.
   n. Truth is mandatory. The smallest lie will be discovered and will be immediately harmful. Remember the public has a right to know the truth.
   o. If the interviewer is hostile, don't assume his or her attitude. Don't get angry or lose your temper.
   p. Do not use, or repeat unverified terminology or "facts" given by a reporter, unless you are positive of their accuracy
   q. **THERE IS NO SUCH THING AS "OFF THE RECORD"**

II. Safety Officer

The Safety Officer is responsible for monitoring; managing and assessing stressful, hazardous, and unsafe situations. Develops measures for assuring personnel safety. The Safety Officer will correct unsafe acts or conditions through the regular line of authority, although the officer may exercise emergency authority to stop or prevent unsafe acts when immediate action is required. The Safety Officer may be sent into the field during rescue operations.

A. Activation
   1. Large search with sufficient hazards
   2. During moderate to complex rescues
   3. Whenever sufficient hazards exist
   4. Whenever IC appoints position

B. Position qualification: IC/IS/Rescue Specialist/specific training.

C. Roles and Responsibilities
    1. Obtain briefing, guidelines, and operating procedures from the IC

> **Safety is everyones' responsibility, not just the safety officer's.**

    2. Advise IC on safety matters
    3. Identify potentially unsafe situations
      a. In types of tasks going out
      b. In air operations
      c. In equipment being used
      d. In fatigue levels of personnel
      e. In stress levels of personnel
      f. In food and water provided
      g. In use of inadequately trained or equipped personnel
    4. Prepare, distribute, and implement preventive information and procedures
    5. Prepare safety message for IAP
      a. Identifies hazards
      b. Lists preventive measures
    6. Follow up on recommendations submitted or problems discovered to ensure risk has been eliminated or reduced
    7. Participate in planning meeting
    8. Review and approve medical, rescue, and evacuation plans
    9. Ensure a communications plan provides coverage to all searchers
    10. Prepare safety briefing to overhead personnel
    11. If required Safety Officer prepares an accident report to IC
      a. Investigate accidents
      b. Assist law enforcement officials
      c. Compile pertinent facts
      d. Works with compensation for injury specialist
      e. Prepares a report to IC
    12. Hazard mitigation
      a. Fatigue
        (1) Help prepare a demobilization plan to ensure searchers have adequate rest before traveling. Make sure sleeping arrangements are secured for any searcher asking for a quiet place to sleep.
        (2) Checks out departing personnel for fatigue
          (a) How many hours since last slept?
          (b) How long did you sleep?
          (c) How long is your drive home?
          (d) Will you have passengers?
          (e) How are you feeling?
          (f) Are you able to stay for additional rest?
          (g) Pull over and take a 20-minute nap if you become sleepy.

    b. Incident Stress
        (1) Take proactive steps to minimize incident stress
          (a) Keep CISD out of field
          (b) Monitor personnel on-scene times at stressful scenes
          (c) Ensure adequate briefing before exposure to stressful scene
          (d) Prepare decontamination equipment and area
        (2) Determine need for critical incident stress debriefing (CISD) for searchers with IC. Work with team liaison officers for teams traveling off site. Ensure follow-up part of demobilization plan.
        (3) Arrange on site critical incident stress defusing

    c. Semi-technical and Technical evacuations
        (1) Participate in and provide field safety control of rescue and evacuations
        (2) Needs to be prepared to go into field with proper safety equipment
        (3) Safety Officer often appointed specifically for evacuations. Reports to the Evacuation Team Leader in a manner similar reporting to the IC back in base
        (4) Oversees all safety aspects of evacuation including personnel and equipment

    d. Terrain/weather hazards
        (1) Works with Plans, OPS , and the IC in identifying when specific terrain or weather conditions may preclude field operations
        (2) Decision factors include
          (a) Training of searchers
          (b) Condition of searchers (fatigue, physical shape)
          (c) Likelihood of subject being alive
          (d) Duration of the search
          (e) Current and expected weather conditions
          (f) Effectiveness of searching (POD) under current weather conditions
          (g) Types of tasks
          (h) Terrain hazards

III. Liaison Officer

The Liaison Officer is the point of contact for the assisting and cooperating agency representatives (AR)
    A. Activation
        1. Activated by IC
        2. Number of agency representatives exceeds IC span of control

    B. Position qualification: IS/ base qualified, diplomatic ability

    C. Roles and Responsibilities
        1. Obtain briefing and report to IC

2. Provide a point of contact for assisting/cooperating with AR
3. Identify Agency Representatives (AR) from each agency, determine methods to contact AR
4. Monitor incident operations to identify current or potential inter-organizational problems

5 minutes of meeting and greeting may help intra-group relationships more than hours of a well-run mission. Good food also helps.

IV. Assistant Liaison Officer (Family Liaison)
The Family Liaison Officer is responsible for providing a point of contact for all family members for information on the status of the search.

A. Activation
1. Many family members
2. Whenever investigation unable to fulfill function
3. Whenever family requires special attention
4. Whenever activated by IC
5. Position can be combined with liaison officer or investigations

A good family liaison will do more to reduce the chance of future civil liability action than the best search strategies and tactics.

B. Position Qualification: IS/experienced FTL/FTM with sensitivity/maturity

C. Roles and Responsibilities
1. Obtain information
   a. Report to and be briefed by IC or liaison officer
   b. Obtain periodic search status updates from IC, OPS, investigations
2. Gather information from the family
   a. May initially fulfill investigative role. As search develops, the role of the investigator and family liaison may conflict.
   b. Interview all family members individually. Analyze differences
   c. Periodically ask questions again
   d. Watch family members discretely for suspicious behavior
      (1) Concealing information, uncooperative
      (2) Lack of concern or emotion
      (3) Suspicions are often incorrect
3. Inform family of search efforts and developments
   a. Explain search strategy and tactics
   b. Explain types of resources being used
   c. Show planning, completed, and in progress maps
   d. Inform family of developments (clues) with discretion
      (1) Consult with law enforcement if criminal possibility
      (2) Avoid destroying verifiability of clues
      (3) Avoid complete disclosure of unevaluated injured or DOA finds

        (4)   Meet with family on a regular basis to keep informed

        (5)   Arrange a tour of base. (Avoid family staying in command post)

        (6)   If requested or deemed appropriate introduce family to IC, investigator, law enforcement personnel, etc.

  4. Comfort family

    a. Explain if volunteers are being used or people are donating services.

    b. Keep an open line of communication

        (1)   Keep informed

        (2)   Give opportunity for questions or input into search

        (3)   Give family a feeling of control

    c. Mentally crutch and prepare family for injury, death, suspension

        (1)   Remind of realistic hopeful options

           (a) Found shelter

           (b) "Bastard" search (use other term)

           (c)   Difficult to find due to mobility or survivable unresponsiveness

           (d)   Reducing probabilities of survival should be brought out as time goes on

        (2)   Prepare for possible injury or death

           (a) Ask appropriate questions about subject's health and equipment

           (b) Explain possible problems of exposure, time

           (c)   Collect information for family plan, media plan. Determine how they would like to be informed if person found DOA

           (d)   Explain evacuation plan, training, and supplies. Discuss the need for some time delay once subject found to evaluate information and organize response

  5. Report to Staff

    a. Prepare a written report at the end of each operational period (unit log may serve purpose). Report provides information on general inferences, tips on dealing with family, suggestions for strategy and tactics, suggestions for follow-up with family.

    b. Update family plan

    c. Provide information to investigator to follow up on

    d. Assist IO in family aspects of media plan, inform medical unit leader of family medical needs for medical plan. Expect grief reactions or possible heart attacks when subject found DOA.

Family Plan	¹ Incident Name	² Date Prepared	³ Time Prepared	⁴ Operational Period

⁵ Family Members ☐ additional sheets

Name	Relationship	Contact Number	or Location

⁶ Procedures if subject found alive and well
☐ Primary Contact(s) _____
☐ Notifier _____

⁷ Procedures if subject(s) found injured
☐ Primary Contact(s) _____
☐ Notifier _____

⁸ Procedures if subject(s) found dead
☐ Primary Contact(s) _____
☐ Notifier _____

⁹ Additional Procedures

Family Plan dbS 303 1/96	¹⁰ Prepared by (Family Liaison Officer)

Fig 10.3 *The process of completing the family plan will help both searchers and the family in the long run.*

UNIT 10: PLANS

I. Introduction

A. IC overview of Plans
1. IC the only person in good position to make sure information exchange between plans, operations, and investigation taking place
2. IC often fulfills Plans role on smaller searches
3. Need to plan is immediate if not required before a search even begins
4. Plans still misunderstood by those not familiar with ICS

B. PLOPS during the first shift
1. Initial arrival to search priority is on getting the teams out
2. Dedicating a highly trained individual to think about 12-24 hours in future is poor allocation of resources
3. PLOPS- Planning and Operations
4. PLOPS- This person <u>is</u> the plans officer and responsible for all aspects of plans
5. Initial responsibilities
   a. Objectives
      (1) written or understood
      (2) determine scenarios and the emphasis of each
   b. Create medical plan
      (1) Who will be the medic?
      (2) What level of medical care is available and how is it obtained?
      (3) Hospitals to take patient to and their level of care
      (4) Aeromedical services available, where and how obtained?
   c. Evacuation plan
      (1) Availability of Stokes basket/patient packaging
   d. Is current strategy in line with planning information (theoretical, statistical, deductive, subjective)
6. Once completed
   a. Shift to tactical planning (writing tasks)
   b. Dual responsibility as deputy Operations

Objectives
- ☐ Understand role of PSC and investigations
- ☐ Understand function of PSC during initial shift and its relationship to OPS
- ☐ Demonstrate the ability to determine search areas using theoretical, statistical, deductive, and subjective process.
- ☐ Demonstrate the ability to determine planning segments, search sectors, and perform Mattson consensus.
- ☐ Demonstrate the ability to make and review an IAP including a 202, 203, 204, 205, Demobilization, SISs, rescue/evac, media, and family plan.

Failure to plan is failure.

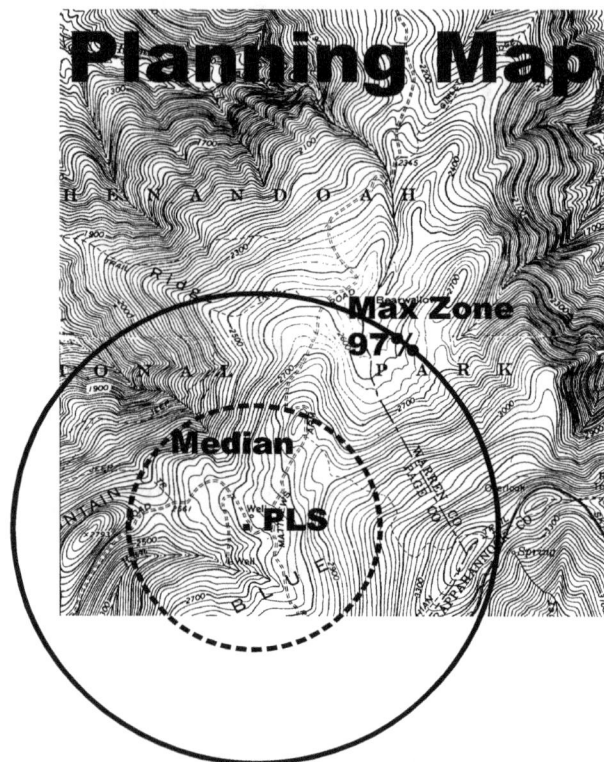

Fig 10.1 *Planning map with PLS, Median distance and Max Zone distances plotted. A common pitfall of many searches is when the search area overlaps several maps. Search planners must not let the edges of a map define the search area.*

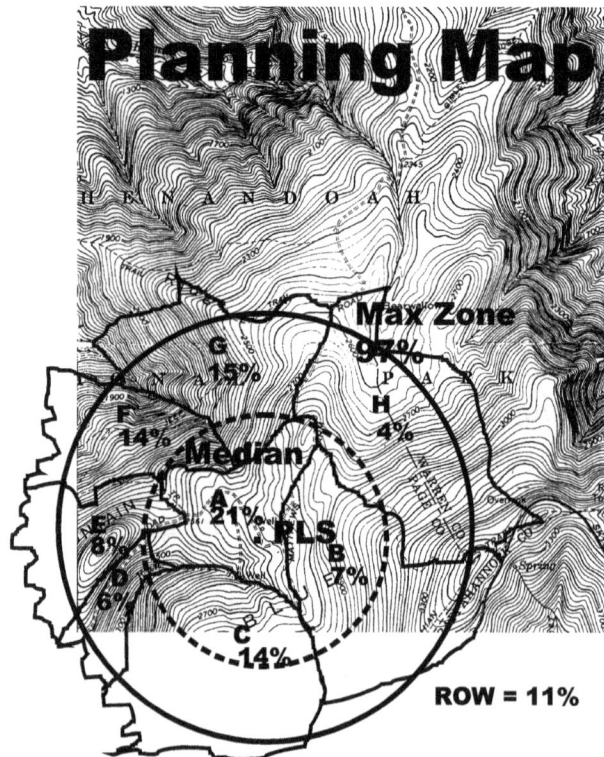

Fig 10.2 *Mattson Map using planning segments. One limitation of most Mattson segments is they don't help define the importance of drainages versus spaces. Some planners will make a fixed distance off of a road, trail, or drainage a planning segment.*

   c. May take responsibility for Clues
7. Still responsible for all functions of plans
8. If deputy OPS assignment is made and PLOPS is not; who is responsible for plans?

C. Use of planning tools
   1. Role of any tool
   2. Theoretical statistical lines
      a. Critical for establishment of search area
      b. Critical to visualize if figures not memorized
      c. Useful for explaining search area to media and family

   3. Planning segments versus Search Sectors

      a. All segments need to be based upon geographical boundaries that a team can potentially find in the field

      b. Planning segments
         (1) Typically 6-12 segments created
         (2) Segments based upon different scenarios
         (3) Number of segments remains constant regardless of the size of the search area
         (4) Purpose of the planning segments is to allow prioritization of scenarios as it relates to potential areas to search within the range of human factors (1-10)
         (5) Allows Mattson to be carried out more efficiently

c. Search sectors
(1) Number of sectors depends upon the size of the search
(2) Size of each sectors depends upon type of terrain and ground cover. If 15-30 minute travel times to task, the size of the area typically allows the team to complete the task in 4 hours. In wooded areas the size often adjusted to 160 acres.
(3) On larger searches the number of search sectors can become large. 1 mile radius = 12 search sectors. 3 mile radius = 113 search sectors 5 mile radius = 314 search

sectors

Fig 10.3 *An incomplete Mattson Map based upon search sectors. A total of 48 sectors are depicted (based upon a goal of 160 acre sectors). Numbers are percentages which have been rounded to the nearest whole number. On larger searches the ability to differentiate between 1-3 percent is difficult.*

(4) Some search planners prefer using search sectors with computer aided systems

4. Mattson
a. Methods
(1) Traditional method covered in MSO/MSF/MLPI
(2) Modified method used in computer systems
(3) Importance of calculating ROW (segment x) first
b. Timing
(1) Mattson should be completed within the first operational period
(2) Mattson may be completed after dispatch of reflex tasks
(3) Holding up dispatch of resources until Mattson completed seen when search planners inexperienced or lack intuition
(4) Important to define search area
c. Updating
(1) If no significant clues shifting POA may be used
(2) If truly significant verifiable clue consider a new Mattson
(3) Some computer-aided system can adjust for significant clues
(4) Expanding or contracting search area may require a new Mattson

     d. Participants
       (1) Most valuable input comes from those with experience but slightly different values
       (2) Including participants solely for political purposes may be required but should be minimized
     e. Most Common Problem
       (1) Mattson calculated by planning staff
       (2) Posted on Mattson or planning map
       (3) Ignored by operations in establishing priority
       (4) Recall original purposes of Mattson

II. Planning Section Chief Role and Responsibilities

The planning section chief is responsible for the evaluation and use of information about the development of the incident and future contingencies. Information is needed to (1) predict probable course of incident events, and (2) prepare alternative strategies and control operations for the incident. Planning focus is on meeting the needs of the next shift.

Question?
After collecting search and planning data how long should it take to determine the:
● Search Area
● Search scenarios
● POA or priority for initial tasks
● Required resources
● Initial tasks

     1. searching data
     2. planning data
     3. political data
     4. need for written IAP
     5. primary goals and objectives

A. Activation
     1. Second shift
     2. Need to create written IAP
     3. Need to update plans
     4. Large search or one rapidly escalating
     5. Need to prepare tasks a shift ahead
     6. Activated by IC

B. Obtain briefing from IC

C. Reports to IC

D. Activate planning section units as needed
     1. Investigations
     2. Technical specialists (weather)
     3. Demobilization
     4. Documentation
     5. Restat
     6. Tactical planning

E. Supervise planning section staff

F. Work closely with all command staff and general staff when preparing IAP
   1. IC- control objectives
   2. OPS- need for divisions, branches. Geographic or function boundaries, types of resources, needed, evacuation, rescue plan
   3. LSC- operational facilities, food, support functions, communications plan, traffic plan
   4. IO- press plan
   5. Safety- medical plan

G. Prepare copies of IAP and determine distribution needs

H. Prepare information of alternatives strategies

I. Identify need for use of specialized resources such as FLIR, night vision, boats, air support

J. Plan for worst case scenarios

K. Ensure shift summary maps completed, sign in sheets completed, staging area manager (or restat) compiling on scene resources and projected resource lists, clue map completed, subject information sheet updated, investigation reports, and any other planning data and logs completed.

L. Prepare recommendation for IC to escalate or systematical shut down search

M. If required brief new planning section chief (typically only one PSC/incident)

III.   Planning Maps

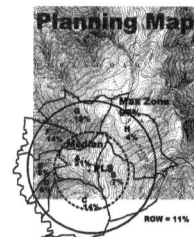

Fig 10.4
Planning Map

A. Planning process map
   1. Contains statistical plots (median, 80%, 90%, etc)
   2. Planning sectors with Mattson scores, possible division locations

B. Tactical planning map
   1. On acetate
   2. Shows possible tasks ideas
   3. Often color coded to resource type
   4. Indicate when a task actually dispatched
      a. remove lines
      b. change color
      c. fill in
   5. Details provided in Chapter 4. (See **Fig 4.8**)

C. Documentation map/Task Completed Map
   1. Hard copy

2. On paper, color coded
3. One/shift
4. Stapled to outside of folder
5. More shifts on acetate overlay with permanent marker
6. Alignment needs

IV.   Incident Action Plans (IAP)

A. Need for written IAP
   1. High turnover rate among staff
   2. Search becoming large or complex
   3. Unified Command needs details written out to avoid confusion among different agencies
   4. IC or PSC feels the need for a written IAP

B. Distribution of the IAP
   1. Copy must go to all Command and General Staff members. Copy must go to documentation for final record
   2. Copy should go to AA/RA, all staff members working in base, and agency representatives
   3. Distribution to Field Team Leaders depends upon the amount of applicable information the IAP contains. Most critical information is captured on the TAF. Any FTL that requests a copy of the IAP should be given a copy. Copies should not be distributed to the press or family. Although parts of the IAP may be shared on a case by case basis.

C. 202 Incident Objectives
   1. Drawn up by Plans reviewed and signed by IC
   2. Shift determined by IC/Plans often divided by day/night OPS
   3. Incident name is subjects last name
   4. Objectives need to be taken seriously, ICS is management by objectives
   5. Good objectives need to met four criteria
      a. Measurable
      b. Veritable
      c. Achievable
      d. Reasonable
   6. Objectives are even broader than strategies

Fig 10.5 *Incident Objectives ICS form 202.*

7. Different camps on objectives
   a. Some believe objectives must apply to an entire search
   b. Other write objectives specific to shift
   c. Some say finding subject is not a good objective (may not be achievable). Other state the only real objective to missing person searches is to locate the subject.
   d. Others include finding the subject on a shift by shift basis
   e. Some use the same four objectives on all ground searches
8. Safety messages/make sure it gets to the FTL

Examples of Different Search Objectives
- Locate subject within operational period
- Increase investigative information
- Provide for searcher safety
- Keep subject within search area
- Search 3 highest POA areas to 80% PODcum
- Increase diversity of resources in high POA areas.
-

INCIDENT OBJECTIVES	1. INCIDENT NAME Shuping	2. DATE PREPARED 18 Dec. 1997	3. TIME PREPARED 00:30

4. OPERATIONAL PERIOD (DATE/TIME) 18 Dec. 1997 06:00 - 18:00

5. GENERAL OBJECTIVES FOR THE INCIDENT (INCLUDE ALTERNATIVES)

1. Locate subject within operational period.
2. Increase investigation to further develop behavioral profile
3. Provide for searcher safety.
4. Keep subject within search area.
5. Search all drainages second time by end of operational period.
6. Search planning segment A,D,F, and G to 75% POD$_{cum}$

6. WEATHER FORECAST FOR OPERATIONAL PERIOD

Clear and Cold. Highs 20F Lows -5F. Winds 10-15 mph. NW

7. GENERAL/SAFETY MESSAGE

Watch for hypothermia. Drink plenty of fluids. Watch for ice.

8. ATTACHMENTS (✓ IF ATTACHED)

☒ ORGANIZATION LIST (ICS-203) ☒ MEDICAL PLAN(ICS-206) ☒ Family Plan
☒ ASSIGNMENT LISTS (ICS-204) ☒ INCIDENT MAP ☒ Media Plan
☒ COMMUNICATIONS PLAN (ICS-205) ☐ TRAFFIC PLAN ☒ Demob/rescue

ICS-202 04-94

9. PREPARED BY (PLANNING SECTION CHIEF) Ken U. Think

10. APPROVED BY (INCIDENT COMMANDER)

Fig 10.6 *Example Incident Objectives on ICS form 202. The 202 also serves as a cover sheet for the rest of the IAP.*

D. 203 Personal Assignments
  1. Staffing of positions one of the most important decisions made by the IC
  2. Plans makes suggestions to fill staffing positions
  3. IC has final approval. Best to work closely with IC
  4. Factors to consider in making assignments discussed in section 2-4
  5. Information that must be collected
     a. Who is here, how long can they stay, how much rest do they need
     b. Who is enroute to the search
     c. Training/position qualification levels
     d. Knowledge of the search
     e. Overall skill/compatibility of the person
     f. Need to integrate multiple agency members into staff
  6. Must obtain projected structure (divisions, groups, etc) from operations and plans. Structures may be limited by the availability of staff.
  7. Monitor Staff to Field Resources ratio

Fig 10.7 *Example ICS 203 Personal Assignment form. Form works best when constant operational periods can be followed. Among volunteer agencies a higher degree of flexibility may be required.*

E. 204 Division Assignments
  1. Division planning considerations
    a. Sufficient Need
      (1) Geographical separation
      (2) Command and Control enhanced
      (3) Communications
    b. Sufficient staff. Division supervisor should be trained to the IC III, IS, GS, or OPS level. Division typically requires one assistant at the FTL level
    c. Sufficient logistical support
      (1) Similar to setting up a search base
      (2) Facility
      (3) Communications
      (4) Transport
      (5) Food/water
      (6) Medical/evacuation support
      (7) Operations kit/forms
    d. Sufficient set-up time
      (1) Best to be set up day ahead of need
      (2) Staff should be in place one hour before field resources arrive
      (3) Initial tasks for initial deployment should arrive with staff

ASSIGNMENT LIST	3. INCIDENT NAME Operation Snowdrift	DATE PREPARED 01/09/96	TIME PREPARED 2015
1. BRANCH I	2. DIVISION/GROUP B		4. OPERATIONAL PERIOD DATE 01/10/96 TIME 0700-1900

5. OPERATIONS PERSONNEL

OPERATIONS CHIEF K J DIVISION/GROUP SUPERVISOR J C

BRANCH DIRECTOR C M AIR OPERATIONS DIRECTOR S. G.

6. RESOURCES ASSIGNED THIS PERIOD

STRIKE TEAM/TASK FORCE RESOURCE DESIGNATOR	LEADER	NUMBER PERSONS	TRANS. NEEDED	DROP OFF PT./TIME	PICK UP PT./TIME
D1 SHP 1150 (Shenandoah Dozer)	Nelson C.	2	No	DP 30 0700 Swift Run Gap	DP 30 1900 Swift Run Gap
Front End Loader with Snowblower	J. C.	1	No	DP 30 0700 Swift Run Gap	DP 30 1900 Swift Run Gap
Virginia Natl. Gd Half track vehicle	Sgt. K/S. B.	4	No	ICP 0700 Park Headquarters	DP 30 1900 Swift Run Gap
Patrol	J P/R. M	2	No	DP 25 0700 Big Meadows	DP 25 1900 Big Meadows

7. TACTICAL OPERATIONS

--Drive to Big Meadows. Clear out occupied Big Meadows residences and emergency equipment bays.
--Begin opening one lane to Rapidan Camps. Evacuate stranded persons to Swift Run Gap. Provide any aid to those persons as needed.
--After the above is accomplished, begin plowing the Skyline Drive north to Skyland, one lane for emergency access only.

8. SPECIAL INSTRUCTIONS

9. DIVISION/GROUP COMMUNICATIONS SUMMARY

FUNCTION		FREQ.	SYSTEM	CHAN	FUNCTION		FREQ.	SYSTEM	CHAN
COMMAND	LOCAL	166.400	SNP	1	SUPPORT	LOCAL	----	----	---
	REPEAT	166.100	SNP	2		REPEAT	----	----	---
DIV./GROUP TACTICAL		----	----	----	GROUND TO AIR			----	---

Fig 10.8 *Division Assignment Form used by the NPS during the Blizzard of 96 (names have been removed).*

    2. Review of forms
      a. Division Assignment form similar in purpose to TAF. Provides information on
        (1) Resource designator
        (2) Personnel involved
        (3) Brief description of tactical assignments
        (4) Communications
      b. Adaptation of the form for other public safety disciplines has occurred.
      c. Form best used only when divisions set up

F. 205 Communications
    1. Immediate Communication Plan
      a. Based largely on pre-plans
      b. Primary SAR frequency for state or county
      c. Should be free of interference at a state/county wide level
      d. Simplifies selection of frequency during searches
      e. Allows teams traveling to search the ability to contact Command Post when in area
    2. Review of Form terminology
    3. Different nets that have been used on searches

Net Type	Function
Tactical	Most common; Base (CP) to teams in field. Several different tactical nets possible: 1) Geographic areas (divisions) 2) Specific functions (dog and foot searchers on different frequencies)
Command	Division to command, used for communication in Command Post, OPS, PSC, Staging to OPS, etc.
Logistics	Food, supplies, Transportation, may be combined with command.
Ground-Air	Command Post to Air resources. A second channel or the ability of air to ground tactical resources is also desired.
Air-Air	Communication between incident Aircraft is required if multiple aircraft in the same air sector.

Table 10.1 *Common Communication Nets used on ground searches.*

    4. Scanning radio help but generally one operator/radio channel being used
    5. Need to inform non-SAR trained personnel about status codes and clear the net
    6. Use of separate frequencies for rescue/evac channel
      a. advantages
      b. disadvantages
    7. Need to gather information from IC, OPS, PSC. Factors to consider in

making plan
   a. Geographic size of search
   b. Topology of search
   c. Have communication holes been identified
   d. Number of people/teams projected
   e. Use of branches/divisions
   f. Trained communication staff available
   g. Equipment available
      (1) Base/mobile radios
      (2) MAST antennae
      (3) Hand held radios
      (4) Repeaters
      (5) Relay sites
      (6) Support equipment
   h. Equipment ordered or projected

> **KISS - Keep is Simple Stupid**

8. On larger searches or if needed consider a briefing by the communications section while picking up equipment as part of the initial briefing process

INCIDENT RADIO COMMUNICATIONS PLAN		1. INCIDENT NAME Shuping	2. DATE/TIME PREPARED 18 Dec. 00:45	3. OPERATIONAL PERIOD DATE/ 06:00 - 18:00	
4. BASIC RADIO CHANNEL UTILIZATION					
SYSTEM/CACHE	CHANNEL	FUNCTION	FREQUENCY	ASSIGNMENT	REMARKS
VASRC BK	BK 3	Tactical (ground teams)	155.205	All ground teams	No PL
	BK 4 DE 1	Tactical (Air-scent/Trailing dogs)	155.160	All dog teams	No PL
	BK 12 DE 10	Backup Tactical	TX 166.300 PL126.5 RX 166.900	All	Back-up channel primary channel fails.
	BK 3	Air-Ground	155.205	VSP helicopter	No PL. Used for intial contact a coordination
ICS-205 04-94	5. PREPARED BY (COMMUNICATIONS UNIT) M. Talk				

Fig 10.9 *Communications Plan.*

9. Selection of Communications Unit Leader and Base Radio Operators
   a. Communication Unit Leader
      (1) Must be able to set up all equipment
      (2) Track the signing in and out of all equipment

(3) Develop communications plan in conjunction with OPS, PSC, and LSC
(4) Follow applicable FCC and incident rules
(5) Track status of teams out in the field
(6) Select and supervise staff
(7) Trouble shoot technical problems
(8) Pass information onto appropriate section

b. Base Radio Operator
(1) Log traffic in communications log
(2) Pass requests, clues to appropriate section
(3) Realize job is to be an organic repeater and not to make operational decisions
(4) Keep calm voice under stressful situations
(5) Voice easy to understand over radio

10. Need to provide Communications Unit Leader with incident maps if not part of the IAP. CUL will typically place maps under acetate to track progress of teams

11. Additional Forms used in Communications
a. Equipment Sign-In
b. Communications Equipment Sign-Out
c. Status Reminder.

EQUIPMENT SIGN-IN	[1] Incident Name Shuping	[2] Time/Date Begun 06:00 17 Dec. 1997	[3] Operational Period	[4] Page _____ of _____ Pages	
[5] Equipment Description	[6] Quantity	[7] Owner (Group)	[8] Contact Person	[9] ETD	[10] Base Location
Bendik-Kings	8	VASRC/CHO	William Dixon	indef	Communications
Motorola 1000	6	VASRC/CHO	William Dixon	indef	"
Wilson Mobile "Base"	2	VASRC/CHO	William Dixon	indef	"
Bendix-Kings	5	VASRC/BKB	Albert Coyne	Dec 19	"
Radio Harness	12	VASRC/CHO	William Dixon	indef	"
Extra batteries (AA)	1 box	VASRC/CHO	"	indef	"

Fig 10.10 *Equipment Sign-Sheet. Form may be used solely for communications gear or for any team gear brought and placed into the search equipment pool.*

Communication's Team Tracking Log	1. Incident Name		2. Date Begun			3. Operational Period	
Team Designator	Task Number	Freq.	Time Start Task	Last Coord.	Time Last Check-In	Est. % Task Completed	Time Back
Notes:							

Fig 10.11 *Communication's Team Tracking Log used by Base Radio Operators to track the last position and contact time of field resources. Field resources should be contacted at least once an hour.*

G. 206 Medical
  1. Creating the immediate medical plan
     a. Determine availability of local rescue squad
     b. Determine if capable of going into field/weather for extended time
     c. Determine location and field packing of medical gear
  2. Initial medical plan often in head of IC/PSC. Safety dictates an initial plan the moment teams are dispatched into the field
  3. Completing the 206 Form
     a. Review of terminology
        (1) Medical Aid
        (2) Ambulance Service
        (3) Incident Ambulances
        (4) Hospitals
     b. Sources of Information

MEDICAL PLAN	1. INCIDENT NAME Shuping	2. DATE PREPARED AND TIME PREPARED 18 Dec. 1997, 2130	4. OPERATIONAL PERIOD 12/18 0600-1800

5. ~~INCIDENT MEDICAL AID STATIONS~~ Helispots (LZ) Coordinates

~~MEDICAL AID STATIONS~~ LZ Descerbtion	LOCATION LORAN	PARAMEDICS	
		YES	NO
Big Meadows	(383970 0782810)		

6. TRANSPORTATION

A. AMBULANCE SERVICES

NAME	ADDRESS		PARAMEDICS	
			YES	NO
Shenandoah National Park	Luray, VA	1-800-555-0911	XX	
Area rescue squads: Luray, Greene, Grottoes, Madison, Madison, Western Arbemarle	(contact park comm center to have dispatched)	1-800-555-0911	XX	
University Aeromedical Rescue	Charlottesville VA	1-800-555-0911	XX	

B. INCIDENT AMBULANCES

NAME	LOCATION	PARAMEDICS	
		YES	NO
NPS Rescue 2	Matthew's Arm Campground	XX	

7. HOSPITALS

NAME	ADDRESS (Numbers are Loran coordinates)	TRAVEL TIME AIR	TRAVEL TIME GROUND	PHONE	HELIPAD YES	NO	BURN CENTER YES	NO
Warren Vet. Memorial (North)	Front Royal, VA (385560 0781170)	25 min	90 min	703-555-0300	XX			XX
Page Memorial (Central)	Luray, VA (383970 0782810)	25 min	75 min	703-555-4561	XX			XX
Rockingham Memorial (Cent)	Harrisonburg, VA (382638 0785768)	20 min	60 min	703-555-8311	XX			XX
Augusta Medical Center (South)	Fishersville, VA (380573 0785907)	20 min	75 min	703-555-4000	XX			XX
Winchester Medical Center (North)	Winchester, VA (391163 0781106)	30 min	120 min	703-555-5198	XX			XX
Univ. of VA Medical Center	Charlottesville, VA (380180 0782960)	10 min	30 min	804-555-2231	XX		XX	

8. MEDICAL EMERGENCY PROCEDURES

Locate and access the patient. Conduct a quick assessment and contact the Shenandoah National Park Communications Center by radio or telephone (1-800-555-0911, 1-540-555-2227, 1-555-999-3422). Remove the patient from danger if necessary. Stabilize the patient. Begin evacuation only if it is in the patient's best interests (do not cause further harm). SNP Comm Center or responding emergency personnel will coordinate evacuation and transport.

ICS-206 04-94	9. PREPARED BY (MEDICAL UNIT LEADER) R. Squaw	10. REVIEWED BY (SAFETY OFFICER) B. Safee

Fig 10.12 *Example Medical Plan with modifications for SAR use.*

   c. Additional information required
      (1) Identifying medics/location
      (2) Keeping highest medic available
      (3) Identifying if using local rescue squad versus SAR medic
      (4) Enforcing medical protocols
      (5) Aeromedical Services
         (a) Helispot (LZ) coordinates
         (b) How to contact Aeromedical service
         (c)   Helispot security and marshals
      (6) Location of nearest Veterinary hospital for dogs and horses
   d. Factors to consider on dedicated SAR standby medical teams
      (1) Availability of resources
      (2) Need to locate subject first
      (3) Relative risk to searchers/subject
      (4) Likelihood of locating subject
      (5) Ability to redirect field resources

Medical Plan Part B	[1] Incident name Shuping	[2] Date prepared 18 Dec. 1997	[3] Time prepared 22:30	[4] Operational period 06:00 - 18:00
[5] Incident Medics				
Name	Medical Training	Organization	Location/Assignment	
Chip Myers	EMT-P	VASRC	Base/Medical Standby	
Darren Chen	EMT-P	VASRC	Field/	
Rita Krenz	EMT-I	VASRC	Field/	
David Stooksbury	Park Medic	NPS	Field/	
Richard Squaw	Park Medic	NPS	Base/Medical Unit Leader	
Bridget Safe	EMT-I	NPS	Command Post/Safety Officer	
Ken Think	EMT-I	NPS	Command Post/PSC	
Ingrid Get	EMT-B	VASRC	Base/LSC	

[6] Medical Equipment (attach additional sheets if required) ❑additional sheets			
Equipment	Current location	Equipment	Current location
Portable O2/extra tanks	Rescue 2		
ALS pack	"		
Basic EMT pack	"		
Pt packaging kit	"		
Hypothermia Kit	"		

[7] Transportation to Medical Site

All equipment on board rescue 2. Upon confirmation of best approach dispatch vehicle and medic to roadhead location. Keys are in van.

Medical Plan-B dbS 305 1/96	[8] Prepared by (Medical unit leader) Richard Squaw	[9] Reviewed by (Safety Officer) B. Safe

Fig 10.13 *Second part of medical plan captures the identification of possible medics, special equipment, and any special procedures.*

H. Rescue/Evacuation Plan
 1. No ICS form available
 2. Basic needs of all evacuations
     a. Who will be in charge
     b. Where are they located
     c. What special equipment is needed
     d. Where is it located
     e. Any special procedures
 3. Evacuations/Rescues can range from simple to complex
 4. Must plan for worst case scenario
 5. Addresses who will be in charge
     a. Need overall scene coordinator
     b. Rescue specialist
     c. Need for safety officer

Rescue/Evacuation Plan	1. Incident Name Shuping	2. Date Prepared 18 Dec. 1997	3. Time prepared 22:00	4. Operational Period 06:00 - 18:00

5. Possible Scenarios

☐ High Angle Rescue ☐ Hi-line ☐ Cave ☐ Multi-casualty #_____
☒ Semi-technical ☐ Extrication ☐ Swiftwater ☐ Other_____

6. Rescue Specialist Qualified	7. Location/Assignment	8. Availability
Chip Myers	Base/Medical Standby	

9. Rescue/Evacuation Equipment (attach additional sheets if required) ☐ additional sheets

Equipment	Location	Owner	Equipment	Location	Owner
Split Stokes	Patrol 53	VASRC			
Pt Packaging Kit	Rescue 2	NPS			
Extra helmets	Patrol 53	"			
Static Rope 2	"	"			
Static rope 3	"	"			
Semi-tech pack	"	"			
Air-Operations kit	Base	NPS			
Extra flagging tape	Base	VASRC			

10. Transportation to Rescue Site.

Equipment loaded into Patrol car 53. Keys are in the vehicle. Extra flagging tape and air ops kept in rescue cache.

11. Rescue/Evacuation Procedures

Dispatch Medical team first. After find location confirmed determine best evacuation route. Determine if any special rigging will be required. Appoint an overall Team leader. Select the rescue specialist to oversee rigging. If possible reroute a nearby team to find site to start flagging a route to nearest roadhead. Dispatch evacuation team with gear. Determine if best to recall a team to base or meet team at road head.

Rescue/Evac plan dbS 305 /96	12. Prepared by (medical unit leader) Richard Squaw

Fig 10.14 *Example Rescue/Evacuation.*

      d. Where the individuals are located

      e. Base

      f. Field

      g. dedicated standby team

         (1) Advantages

         (2) Disadvantages

         (3) Circumstances that favor dedicated teams

            (a) Multi-casualty scenario with likelihood of survival

            (b) High risk to searchers

            (c)   Location of subject known or find seem eminent

6. Plan what equipment needed

7. Plan where equipment located and how transported to site

I. Demobilization

Fig 10.15 *ICS demobilization form. The form is best suited for checking out a single individual through a rigid process.*

1. ICS form not applicable in a typical volunteer setting

2. Demobilization are rather similar incident-to-incident. A good plan identifies items unique to a particular mission

3. General Outline

    a. Stop incoming resources if appropriate (DES, team Dispatch)

    b. Remove Field teams from field back to (Base, division)

    c. Shutdown Divisions, Relays, Repeaters

    d. Decide if a final mass debriefing will be held/ announce/ hold

    e. Provide logistical support required for teams returning from field

       (1) food, water

       (2) sanitation

       (3) showers, rest, emotional support

    f. Contact logistical sources to cancel or minimize future logistical requests. (Need list of contacts)

    g. Decide to hold staff debriefing

    h. Release volunteers

    i. Prepare documentation

    j. Appoint SAR group coordinators for packing away equipment

    k. Release SAR providers

    l. Clean up base

    m. Release Staff

J. Family Plan
1. No ICS form exists
2. Purpose of form is to facilitate communication with family and determine procedures before a find is made
3. Form should be completed by family liaison after a relationship is developed

Demobilization Plan	1. Incident Name Shuping	2. Date Prepared 18 Dec. 1997	3. Time Prepared 00:50	4. Operational Period 06:00 - 18:00
5. **Contact Organizations to Stop/Adjust Incoming Resources** ❏ additional sheets				
Organization	Resource Type	Contact #	Contact Name	Stop time
NPS	Search resource	800 555-1234	Dispatcher	
VASRC	Search resource	800 555-2772	Dispatcher	
VADES	Coordination for SAR	800 555-8892	SAR Watch Officer	
MCVRS	Local rescue squad	540 555-2167	S.O. Dispatcher	

6. **Release Priorities (organizations, logistics, etc)**

7. **Equipment to Return** ❏ additional sheets			
Equipment/item	Owner	Contact #	Release Priority
Porta-Johns	Rent-a-Castle	703 555-1145	
Generator	MCVRS	540 555-2167	
Copier	MCVRS	540 555-2167	

8. **Personnel Release Procedures**				
Field Resource Removal	Field Facilities Support	Incident Debriefing Time:	Base Breakdown	Safety Check

9. **Additional Procedures**
If CISD required contact number for on-scene response 1-800- 555-0129 Fatigue policy states all drivers must go through check-out procedure, be given MSLT test.

Demobilization dbS 304 4/96	10. Prepared by (Demobilization Unit Leader Ken Think

Fig 10.16 *Example of a demobilization check-out sheet more specific to SAR needs.*

4. Important to have sensitivity in completing information for form
5. Rapid implementation of family plan required
    a. If family in base can detect change in mood or overhear information usually within 5 minutes of find

Family Plan	¹ Incident Name Shuping	² Date Prepared 17 Dec. 1997	³ Time Prepared 23:30	⁴ Operational Period 06:00 - 18:00

⁵ Family Members ☐ additional sheets

Name	Relationship	Contact Number	or Location
T. Shuping	Wife		
B. Shuping	Son		
M. Shuping	Son		
M. Strutt	Friend		
F. Kovach	Friend		

⁶ Procedures if subject found alive and well

☐ Primary Contact(s)__T. Shuping_____
☐ Notifier _____Anyone_____

⁷ Procedures if subject(s) found injured

☐ Primary Contact(s)_____M. Strutt_____
☐ Notifier _____B. Nice_____

⁸ Procedures if subject(s) found dead

☐ Primary Contact(s) _____M. Strutt_____
☐ Notifier _____B. Nice_____

⁹ Additional Procedures

Provide tour of base on a daily basis.
Give air-scent dog demo at 16:00

Family Plan dbS 303 1/96	¹⁰ Prepared by (Family Liaison Officer) Bea Nice

Fig 10.17 *Example family plan.*

b. Need to confirm finds quickly and within reason
    (1) Incorrect finds due to a lost team member (often unknown to base) being reported over the radio as "we found him"
    (2) Sweeps of hospital or institution grounds results in multiple bodies being found. Clothing, age of body provides clear clues
    (3) If uncertain best to inform family of changing developments

K. Media Plan
    1. Not an ICS form
    2. Sources of information
        a. Media: Determine deadlines and any special requests

        b. AA/IC: Willingness to make press interviews
        c. Family (Family liaison officer): Desire to be protected from press
        d. LSC: Site to hold press briefings
        e. OPS: Security and safety requirements that would preclude media
        f. Planning meeting good opportunity to collect information
    3. Major Purpose
        a. Help media meet deadlines
        b. Schedule press briefings
        c. Identify media contacts
        d. Determine Response if find occurs

Media Plan	1. Incident name Shuping	2. Date prepared Dec. 18 1997	3. Time prepared 01:00	4. Operational Period 06:00 - 18:00
5. **Media Contacts** ❏ additional sheets			
6. Media Organizations	7. Contact person	8. Contact number	9. Deadlines	
WMAL (Radio)	Sue Wave	555-3656	11:30, 16:30, 22:30	
Daily Info (Newspaper)	W. Pennington	555-2422	18:00	
WTVN (TV)	John Micheal	555-1324	11:15, 16:15, 22:15	
10. **Regular Briefings**			
Times	Participants	Location(s)	Preparations	
As needed				
11. **Media Procedures for Subject(s) found alive and well**			
Confirm correct subject. 2) Contact family first, family has given permission for a happy reunion photo op. 3) Prepare statement for press. 4) Facillitate reuinion photo op. 5) Schedule summary press briefing. Sheriff wants to be present			
12. **Media Procedures for Subject(s) found injured**			
1) Confirm correct subject. 2) Contact family first, family asks to be protected from media in case of injury. 3) Prepare a statement for the press given by IO. 4) Prepare roadhead photo opp press area try to keep media from field. 5) Schedule summary press briefing with Sheriff and IC.			
13. **Media Procedures for Subject(s) found deceased**			
Same as above			
14. **Additional Procedures**			
Use picnic pavillion as press briefing area in case of rain.			
Media plan dbS 302 1/96	15. Prepared by (IO) Julie Talkwell			

Fig 10.18 *Example of a media plan.*

L. Subject Information Sheet (SIS)
   1. Generate SIS within first 15 minutes. Helps get teams out faster
   2. Investigation supplies information
   3. Other sources of information for preparing the SIS
      a. Initial report completed by first officer on-scene
      b. Lost Person Report (LPR) form
      c. Direct interview
   4. Preparing the SIS
      a. Photo with date when taken, note any changes. If possible see if photoshop will donate services to make color copies
      b. Name and physical description
      c. Items carried description (pocket or backpack items)

Subject Information Sheet

Name: Ken Shuping

Subject Description:

Age: 72   Height: Weight:

Sex: Build: Eyes:

Race: Hair:

Identifying marks:

Place the best photograph of the subject in this space.

Clothing Description:

Date of Photograph:
Changes since taken:

Equipment/discardables:

Search History:

(PLS history and details, subject catagory)

Track information if avialable may be placed in this space

Searcher Information:

(Name to call, evasiveness, etc)

Hazards:

Fig 10.19 *Example pre-formatted subject information sheet (SIS). Many SIS are placed upon a blank sheet of paper with the photograph. The advent of color copies makes it possible to pass-out quality flyers.*

       d. PLS, LKP details

       e. Brief subject behavioral profile

       f. Foot print information

       g. Search hazards in area

       h. Name of person who last created/updated form

       i. Date/time/version number of sheet

       j. IC/RA/AA approval before publishing

       k. Limited publishing to press (item left off or color left out)

          (1) Press usually cooperative in leaving one detail out of broadcast.

          (2) Helps with verification of citizen reported clues

          (3) Best to expect SIS will fall in the hands of the press

          (4) Confidential information should be left off form

M. Planning Map often included

    1. Map should show CP, PLS, Staging areas, Divisions, overall search area

    2. More extensive package put together for divisions/groups

    3. Several different maps may be included. Topographical map, hiking trail map, urban road map, plot maps

N. Base Map

    1. Shows layout of base/command post

    2. Helps crowd control

    3. Consult MSO/MSF/MLPI for base examples

O. Traffic Map

    1. Shows traffic flow patterns around base camp

    2. Often included with base map

    3. Parking problems addressed

P. Unit Log

    1. Common practice to include one copy of unit log at end of IAP

    2. Allows all mission staff to have copy of unit log

Q. Cover Sheet

    1. Practice at wildland fires and other large incidents to place a humorous cover sheet at the first of the IAP

    2. Cover sheet includes incident name, and operational period

R. Safety Message

    1. If not enough space on 202 for safety message a separate sheet may be included

    2. Important for complex or highly hazardous missions

S. IC Review of the IAP

    1. Important to read entire document before signing

    2. Look for completeness

    3. Consistency among the different plans

4. Each plan can be carried-out with equipment and staff available.
5. Ask planning staff to provide alternative plans
6. Keep plan as simple as possible (KISS principle)

V. Investigation

A. Role   The investigation unit leader is responsible for ongoing investigation of the lost subject and all circumstances of the disappearance

1. Activation: all searches

2. Meet with previous investigator or responsible agent to determine what investigation has been completed and by who
   a. who have they interviewed
   b. what have they determined
   c. what are the possible scenarios
   d. what have they done
   e. when where the interviews done
   f. what documentation have they collected
   g. what are they planning to do
   h. have they
      (1) issued an All Points Bulletin (APB)
      (2) checked criminal record
      (3) checked subjects bank account, credit cards
      (4) checked hospitals (both ER and psychiatric ward), doctor offices, clinics, morgue
      (5) entered subject into National Crime Information System (NCIS)
      (6) contacted local and state jurisdictions
   i. check bastard scenario
      (1) residence thoroughly searched
      (2) airports
      (3) bus depot
      (4) hotels/motels
      (5) other relatives, friends, lovers
      (6) bars and other hangouts
   j. Determine how to best work together and how you can help

3. Review or complete Lost Person Questionnaire (LPQ)

4. Review or complete Subject Information Sheet

5. Obtain photograph and/or video of subject

6. Review clue map and log and consult with clue director on a regular basis

7. Review communications log every 2-3 hours to determine if any missed

clues or investigative leads

8. Meet with OPS, planning chief, IC, determine any investigative needs

9. Meet with RA investigator on a regular basis to exchange notes, information, and ideas

10. Report significant leads to IC, OPS, PSC, and clue director

11. Conduct on-going investigation

12. Meet with family members and other sources of information as needed

13. Maintain unit log

14. On multi-shift searches summarize each shifts investigations including additional informational needs

15. Report at planning meeting results of clue follow ups

16. Generate investigative time line. Should start in far past covering significant life events moving to events that lead up to event.

17. Generate behavioral profile specific for lost subject

B. Documentation
   1. Create investigative folder
   2. Label outside of folder with contents
   3. Folder stays with investigator at all times
   4. Duplicates may be copied and sent to documentation unit leader

VI. Documentation
   A. Role
      1. Aids communication among functions
      2. Provides continuity when personnel change
      3. Avoids duplication of efforts
      4. Aids future statistical studies
      5. Assists with mission critique, training materials
      6. Helps fundraising and budget justification
      7. Legal concerns

   B. End of Operational Period Folder
      1. Chronological Logs
         a. Communications Logs
         b. Unit logs of Key personnel. Original Logs may be kept by officers until end of search or until replaced

2.   Task Completed Folder
a. Copy of Task Completed Map (if single task map) stapled to the front
b. Task Unit Log stapled to the inside of the folder
c. Task Assignment Forms placed in order of task number
3.   Registers
a. Sign in logs
b. Equipment logs
4.   Maps
a. Generally posted or moved to Planning section to determine areas covered
b. Hazard Map posted throughout search
c. Commo dead zones, relays and repeaters posted throughout search
d.   Mattsons posted throughout search
5.   Each operational period placed into a box or crate. Must be stored in a secure place with key personnel having easy access to information.

C. End of Mission
1.   Document medical record, scene description, and disturbances
2.   Interviews
a. Interview the subject if possible
b. Interview the find and evacuation team. May be included on debriefing form
3.   Document rationale for suspension. Must be well documented in IC log even if log contains no other information. Should contain information addressing
a. Coverage, probabilities, size of search area
b. Clue follow-up, investigation, indications out of area
c. Resource availability
4.   Copy documentation that needs to be shared or summarized for RA/AA other agencies. Typical critical information to copy
a. Investigative folder
b. Cumulative POD maps
c. Task Completed Maps
d.   Clue map and log
e. Shift Summary Form
f. Mission Summary Form (After Action Report)
5.   Order and label files for long term storage
6.   Make and forward copies to any state agencies if required
7.   Pack and send documentation to long term storage site

VII.   Tactical Planning
A. Purpose: Creation of tactical field assignments for the next operational period
B. Allows operational staff to concentrate on deployment of resources
C. Technique covered in Operations Section

VIII. Technical Specialist
  A. Most common position is a weather specialist
  B. Need to analyze local wind currents for dogs
  C. Need to document all weather since time of last seen for
    1. Determining survivability
    2. Determining aging factors for trackers

IX. Example of Incident Objectives
  A. Situation: Several feet of snow falls during the blizzard of 1996. Several park employees are stranded. Several hikers and campers have been stranded. Normal plow operations are not possible
  B. Objectives with resulting Strategies
  C. Strategic Planning Assumptions
  D. Analysis of Alternative Strategies for Objective Three, Search and Rescue

01/13/96   **Operation Snowdrift**

                    INCIDENT OBJECTIVES AND STRATEGIES

1. Provide emergency access to park residents stranded by snow and provide alternative accommodations for them if needed.

    --Use park equipment to push open one lane to residential areas.
    --House and feed displaced residents at the Ramada Inn in Luray, VA.
    --Strategically pre-position equipment and operators based upon anticipated and actual precipitation, snow/ice removal needs and priorities.

2. Make reasonable efforts to protect high-risk park structures and other infrastructure.

    --As facilities and areas become accessible, identify structures/facilities at risk, and, if needed, plan and perform safe mitigating actions

3. Identify and locate stranded park users, determine their situation, stabilize their situation, and safely extricate them by means appropriate to the situation and conditions.

    --Gather information, investigate clues, assess data and assumptions and analyze the potential and urgency of SAR situations.
    --If information received indicates an urgent SAR problem, rapidly assess the situation and safely respond in an appropriate manner (considerations: locate, assess situation, stabilize people of their situation, evacuate).

    (Note: refer to alternative strategy analysis and selection, attached.)

4. Provide for the basic welfare and storm-related emergency needs of the employees of Shenandoah National Park

    --Operations will conduct sufficient plowing to allow other sections, especially Logistics, to function.
    --Evaluate CIS needs and develop and execute a CIS plan if needed.
    --Provide all reasonable, legal assistance possible to the involved family.
    --Develop, have approved by the Agency Administrator, and execute a plan for park/incident involvement in the funeral of ---------------..

5. Where practical, provide assistance to neighboring agencies.

    --Assess requests as they are received and respond appropriately.

6. Restore the Park's district-based initial attack search-and-rescue and emergency service capability.

    --Plow to provide reasonable, safe employee access to district SAR and emergency caches and critical supporting facilities.
    --Reestablish required occupancy.
    --Return park and district SAR and emergency service supplies and equipment to a response-ready condition.
    --When plowing, avoid unintentionally improving public access until the park is in a response-ready posture.

STRATEGIC PLANNING ASSUMPTIONS

Although no one is known to be stranded in or near the Park, this table outlines the likely situation for any unknown persons who may be stranded in the park on 01/13/96, assuming that they became stranded on or about 01/07/96.

Factor	Rational AT Thru Hiker	Rational Weekend Hiker/Camper	Person with Diminished Reasoning
Food	Could be okay. Could be starting to become low.	Probably in trouble	In trouble
Water	If stationary, probably okay. If moving, may be dehydrated.	If stationary, probably okay. If moving, may be dehydrated.	In trouble
Shelter	May be okay in hut or shelter; but, if hiking without a tent, in trouble	Probably has a tent	In trouble
Warmth	Probably okay.	May be in trouble.	In trouble
Experience	Probably greater	Varies	Irrelevant
Equipment	Probably better	Varies	May be irrelevant
Probability of poor survival decisions	Lower	Higher	Highest
Probability of rendering aid in time because of conditions	High, despite conditions. Subjects are fairly self-sustaining.	Low unless they take measures to make themselves known	Lowest
Probability of being reported by others	Given the time that has passed and the national media coverage, probably high.	High	High
Investigation	Info available to account for ND sightings. Some pre-storm and early incident SD sightings not accounted for, but no family/friends have called.		
Aerial Recon POD$_{CUM}$	Acceptable		
Analysis	Probably okay if prepared. Most likely to be helped by the intervention that can be reasonably provided given the situation and conditions	Probably in trouble. Additional intervention not likely to be in time because of conditions. *Note:* Incident has rescued all known persons.	In trouble. Incident intervention not likely to be in time because of conditions

01/13/96
ANALYSIS OF ALTERNATIVE STRATEGIES FOR OBJECTIVE 3,

SEARCH AND RESCUE

Factor	Strategy A	Strategy B	Strategy C	Strategy D
Risk to Personnel and Equipment	Low	High	High	Moderate
Cost	Low	High	High	High
Time	Longer	Moderately Longer	Longer	Shorter
POS* for anyone out there	Moderate	Moderate	Higher	Moderate
Effects on meeting other objectives	Low	Very High	Moderate	Low

* Success = discovering person and rescuing them in time to ensure their long term survival.

Strategy A = Continued information gathering, investigation and analysis and field actions only occur upon the development of urgent situations.

Strategy B = Continued urgent plowing of one lane for emergency access and facilitating of self-rescue. One lane plowed from Piney River (MP 22) to Blackrock Gap (MP 87).   This mode of plowing is exhausting for both personnel and equipment.

Strategy C = Begin an oversnow check of the unplowed portions of the Drive and related nearby huts.

Strategy D = Additional aerial reconnaissance with ground follow up as needed.

Several combinations of the above alternatives were also considered, but the factor analysis always led to the most extreme or "worst-case" description.

SELECTED STRATEGY: Alternative A.

RATIONALE FOR SELECTION:

1.   Currently, there are no known reports of unknown missing persons in the Commonwealth of Virginia.

2.   The scope of the storm and its aftermath, along with the wide scale national media coverage it was received, makes it likely that any family, friend or acquaintance of someone stranded in the park would be concerned.

3.   Given the time that has passed since the storm ended and the media attention began, it is likely that anyone concerned about

someone possibly missing in the park would have made contact by this time.

4.   None of the unresolved anecdotal reports of people hiking in the park from 01/04 through 01/07 include reports on lone individuals. Two or more people traveling together increase the likelihood of notification and self-rescue.

5.   Overflights were conducted every day, three days in a row, using a variety of observers.

6.   The people who were rescued had no information about others needing help.

7.   There were two known cases of self-rescue, proving that it could be done.

8.   The weather for late Monday through Thursday was clear and relatively warm.

9.   The checks of boundary areas that have been conducted located no abandoned cars in high-use areas where one could reasonably expect to find use in almost all types of weather. These checks have also not found any clues such as tracks, etc.

Prepared by: _____   _____
                Planning Section Chief   Date

Approved by: _____   _____
                Incident Commander   Date

UNIT 11: LOGISTICS

I. IC Overview of Logistics

A. On smaller searches IC often performs role of Logistics Section Chief (LSC). Selection of Unit leaders who are able to function nearly independently makes this possible

B. Outside of the all-volunteer/donation searches, LSC must coordinate closely with FSC and IC to determine procurement policy

C. IC must ensure LSC or unit leaders taking a proactive approach

D. Logistics must work closely with OPS to make sure all operational objectives can be met

Objectives:
☐ Describe the roles and responsibilities of the Logistics Section Chief and staff.

Without logistical support few if any operational objectives can be met.

Selection of the Command Post location and layout may be one of the most important decisions the IC and LSC make.

II. Logistics Section Chief (LSC)

A. Activation
   1. Activated by IC
   2. Incident has potential of going beyond one operational period

B. Position qualification: On large searches IC, IS, base qualified, specific training in logistics. On smaller searches where only food, facilities, transportation, and rescue squad on standby required may use local resource person with authority and capability to provide logistical support. This method results in the communications unit leader reporting directly to IC or OPS.

C. Position functions
   1. Obtain briefing from and report to IC
      a. Subject information
      b. Current resources present
      c. Number of people (present and expected)
      d. Types of resources (present and expected)
      e. Location of base, staging area, other facilities
      f. Other facilities nearby
      g. Current logistical problems
      h. Number and type of vehicles present and needed
      i. Special equipment present and needed
      j. IC needs and desires

2. Activate needed units of logistics
   a. Service branch director
   b. Support branch director
   c. Communications unit leader
   d. Medical unit leader
   e. Food unit leader
   f. Equipment unit leader
   g. Facilities unit leader
   h. Ground support unit leader

3. Establish work location and preliminary tasks for section personnel
4. Identify service and support requirements for planned and expected operations
5. Review IC requirements for next operational period
6. Receive briefing from OPS and PSC
   a. Determine what equipment, resources present
   b. Determine immediate and projected needs
   c. Determine what facilities are/will be required
   d. Determine resources enroute
   e. Determine what planned for next 24 hours

7. Receive briefing of needs from Agency Representatives concerning projections of personnel and equipment if not provided by OPS and PSC
   a. Develop a needs list
   b. Prioritize the needs list
   c. Have IC review priority list

8. Coordinate and process requests for additional logistical support
   a. Document all requests
   b. Determine priority of request
   c. Determine quickest and most cost-effective method of obtaining request
   d. Fulfill request
   e. Document name, address, special requirements of supplier
   f. Furnish item to requesting section

9. Participate in preparation of IAP
   a. Provide input to section leaders and review communication plan, medical support plan, facility, and traffic plan
   b. Review IAP and estimate section needs for next operational period
   c. Determine if divisions will be created
      (1) communications
      (2) transportation
      (3) food, water
      (4) sanitation
      (5) staging area
      (6) division or branch facilities

d. Search area
    (1) expansion, contraction
    (2) areas of concentrated resources
    (3) special equipment requirements
    (4) transportation
    (5) communications
e. Expected personnel
    (1) increase/decrease
    (2) levels of training and type
    (3) time expected on scene
    (4) meal requirements
    (5) housing requirements
    (6) sanitation requirements
    (7) special equipment requirements
    (8) transportation requirements

> **Proactive attitudes and the ability to find anything anywhere are the hallmarks of a valuable logistics section chief.**

10. Advise on current service and support capabilities
  a. Attend planning meetings
  b. Advise IC of any potential shortage that affects operations
  c. Brief staff on the following aspects
    (1) Food plans
    (2) Transportation plan
    (3) Facilities plans
    (4) Communications plan
    (5) Traffic plans
    (6) Equipment plan
    (7) Medical plan
    (8) Procurement procedures
    (9) Lodging plans
    (10) Sleep/nap areas
    (11) Supply needs

  d. Estimate future service and support capabilities
  e. Obtain estimate from IC, PSC, OPS on future numbers of personnel, time on scene
  f. Determine any special requirements or limitations from AA or SCO
  g. Discuss discrepancies in any estimates with IC
  h. Determine need to increase or decrease
    (1) Meals, lodging berths
    (2) Transportation
    (3) Radios
    (4) Other communications requirements
    (5) Use of facilities
    (6) Use of specialized equipment
    (7) Other

11. Receive Demobilization plan from Planning section
    a. Review plan to determine logistical role and needs
    b. Determine personnel requirements when plan implemented
    c. Maintain a current list of all people, organizations with listed contact person and number to inform when find or suspension occurs
    d. Determine plan to stop or reduce incoming flow of materials, supplies, people when demobilization begins

12. Ensure general welfare and safety of Logistics section personnel
13. Maintain the following documentation
    a. Unit log
    b. List of group/personal equipment brought to incident

III. Service Branch Director

The Service Branch Director, when activated, is responsible for the management of all service activities at the incident. The branch director supervises the operations of the communications, medical, and food units.

A. Activation
   1. When span of control of LSC exceeded. Activation of Service Branch Director will only occur on the largest, most complex ground searches. Under most circumstances the LSC can control all units
   2. When LSC activates

B. Position qualification: base qualified, logistical training or authority

C. Position Description
   1. Obtain briefing and report to LSC
   2. Participate in planning meetings of LS personnel
   3. Review IAP
   4. Organize and prepare assignment for Branch personnel (Medical and Food). Communications unit may be placed under service branch director on larger searches
   5. Coordinate activities of branch units
   6. Resolve Service Branch problems

IV. Support Branch Director

The Support Branch Director (SBD) when activated, is responsible for the development and implementation of logistical plans. The SBD supervises the operations of the equipment, facilities and ground support units.

A. Activation
   1. When span of control of LSC exceeded
   2. When LSC activates. Activation of Service Branch Director will only occur on the largest, most complex ground searches. Under most circumstances the LSC can control all units.

B. Position qualification: base qualified, logistical training or authority

C. Position Description
1. Obtain briefing and report to LSC
2. Participate in planning meeting of logistics personnel
3. Review IAP
4. Organize and prepare assignments for Branch personnel (equipment, facilities, and ground support)
5. Coordinate activities of branch units
6. Resolve Service Branch problems
7. Work closely with operations section fulfilling requirements and solving problems

V. Communications Unit Leader

The Communications Unit Leader (CUL) is responsible for developing plans for the effective use of incident communications equipment and facilities, installing and testing of communication equipment, supervision of incident communications, distribution of communications equipment, documentation of incident communications, and the maintenance of communications equipment. On some smaller searches the CUL may report directly to OPS.

A. Activation
1. Activated by LSC or OPS
2. Activated whenever field teams are deployed
3. Activated whenever documentation of communications (phone, radio, etc.) is required

B. Position qualification: IS, CUL qualified, base qualified, CAP qualified, HAM operator

C. Position Description
1. Obtain briefing from Service Branch Director, LSC, or OPS depending upon IC request
2. Determine unit's personnel requirements. (base radio operators, equipment check-out/in, equipment maintenance, telephone operators, messengers, relays, etc.)
3. Supervise staff
4. Advise OPS, PLANS on communications capabilities/limitations
    a. Range of base radio network
    b. Numbers of hand held radios available
    c. Number of frequencies available
    d. Other stations sharing frequencies
    e. Potential radio dead zones
    f. Aircraft frequencies
    g. Telephone lines, need for more
    h. Cellular phones available, needed

INCIDENT RADIO COMMUNICATIONS PLAN		1. INCIDENT NAME Shuping	2. DATE/TIME PREPARED 18 Dec. 00:45	3. OPERATIONAL PERIOD DATE/ 06:00 - 18:00	
4. BASIC RADIO CHANNEL UTILIZATION					
SYSTEM/CACHE	CHANNEL	FUNCTION	FREQUENCY	ASSIGNMENT	REMARKS
VABRC BK	BK 3	Tactical (ground teams)	155.205	All ground teams	No PL
	BK 4 DR 1	Tactical (Air-scent/Trailing dogs)	155.160	All dog teams	No PL
	BK 12 DR 10	Backup Tactical	TR 166.300 PL138.5 RX 165.900	All	Back-up channel primary channel fails.
	BK 5	Air-Ground	155.205	VSP helicopter	No PL. Used for initial contact & coordination
ICS-205 04-94	5. PREPARED BY (COMMUNICATIONS UNIT) M. Talk				

5. Prepare and implement the Incident Radio Communications Plan (ICS form 205) **see page 10-10**
   a. Determine frequencies available on equipment
   b. Determine which frequencies are clear
   c. Determine projected number of teams
   d. Determine if divisions will be used
   e. Determine distance of divisions from base
   f. Number and capabilities of handhelds
   g. Determine need for relays, repeaters
   h. Consider need for separate frequencies for command, logistical, division, aircraft OPS, evacuation, medical, etc.
   i. Consider radio dead zones
   j. Need for additional telephone lines, where they should be placed
   k. Cellular phones available, needed, and restrictions on use

6. Establish work area in close proximity to OPS
7. Establish message center, runners if required
8. Ensure base radio system installed and tested
9. Ensure radio system accountability system established
   a. Sign-in all group/SAR agency radios
   b. Ensure radio, extra batteries, antennae, and harnesses signed out to FTL. Record name, team identifier, task number
   c. Distribute and document radios to divisions, based upon communication plan

10. Ensure portable radio equipment from cache is distributed following communication plan
11. Ensure appropriate section of TAF filled in
12. Make sure check-in procedures followed
13. Enforce radio SOP's
    a. Clear text communications (swear words forbidden)
    b. Concise communications
    c. Mission relevant communications (avoid use of trade names)
    d. Team-to-team communication must be approved by base radio operator
    e. Use of FCC type accepted equipment and wattage.
14. Maintain communication logs
    a. Telephone logs
    b. Radio log
    c. Message log
    d. Equipment logs

15. Ensure equipment is tested upon returning
16. Recover equipment from relieved or released units
17. Ensure important messages documented and immediately passed on to appropriate personnel (Logistics, OPS, clue supervisor, investigation, IC, etc.)
18. Review communications logs at least once an hour and ensure appropriate action taken on all messages
19. Ensure relief of communication personnel (especially remote relays)
20. Ensure communication equipment properly packaged, accounted, and returned to appropriate agency representative or group representative

VI. Medical Unit Leader

The Medical Unit Leader is responsible for the development of the medical emergency plan, obtaining medical aid and transportation for injured and ill incident personnel, and preparation of reports and records.

A. Activation
   1. Whenever searchers are deployed into hazardous conditions
   2. Whenever search subject may require immediate care
   3. By LSC, or SBD, whenever possible
   4. If subject found control of field medical resources switches to OPS

B. Position qualification: EMS officer, EMT or greater, RS

C. Position Description
   1. Obtain briefing from SBD or LSC
   2. Participate in logistics planning meeting
   3. Determine level and number of emergency medical activities performed prior to activation of medical unit
   4. Prepare Medical emergency plan (ICS Form 206). **See page 10-13.**
   5. Plan and prepare procedures and equipment for medical emergency.
   6. Activate medical unit
   7. Respond to request for medical aid and transportation
   8. Respond to request for medical supplies

VII.   Food Unit Leader

The Food Unit Leader is responsible for determining feeding requirements at all incident facilities, menu planning, determining cooking facilities required, food preparation, serving, providing potable water, and general upkeep of the food service area.

A. Activation
   1. Activated by LSC or SBD
   2. Activated when search if projected to last longer than 4 hours

B. Position qualification: Ability to plan and provide food to large numbers of people

C. Position Description
   1. Obtain briefing from SBD or LSC
   2. Determine work area, serving area, number of personnel requiring meals. Obtain numbers of personnel on-scene, projected personnel, family, and press.
   3. Determine best method of feeding both base and field personnel
   4. Set up food unit equipment
   5. Document all organization and companies donating food. Provide copies of list to IC, IO as requested
   6. Ensure all appropriate health and safety measures taken
   7. Ensure sufficient fluids available to meet incident needs. Review fluid needs with safety officer or medical unit leader
   8. Ensure menu meets the high carbohydrate demands of field personnel. Review menu with safety officer or medical unit leader
   9. Supervise food unit personnel
   10. Ensure off-site personnel receive meals
   11. Contact appropriate agencies when demobilization initialized
   12. Demobilize food unit in accordance with demobilization plan

A good meal will improve morale of field resources more than any speech.

VIII. Supply Unit Leader

The Supply Unit Leader is responsible for receiving and storing all equipment and supplies for the incident, maintaining an inventory of supplies, and returning supplies to the appropriate agencies.

A. Activation: Activated by SBD or LSC, typically on larger searches

B. Position qualification: Ability to obtain and manage SAR supplies

C. Position Description
1. Participate in logistics planning meetings
2. Determine the type and number of supplies required by command, logistics, operations, plans, and finance
3. Determine the type and amount of supplies enroute
4. Arrange for receiving ordered supplies
5. review IAP for information on projected supply requirements
6. Order, receive, distribute, and store supplies and equipment. Provide for security of supplies.
7. Receive and respond to request for personnel, supplies, and equipment
8. Maintain inventory of supplies and equipment. Maintain list of suppliers and those making donations.
9. Upon demobilization, determine how to return donated supplies (if required) to suppliers

IX. Facilities Unit Leader

The Facilities Unit Leader is primarily responsible for the layout and activation of incident facilities. The unit provides sleeping and sanitation facilities for incident personnel and manages base and camp operations.

A. Activation: Activated by Support Branch Director or LSC

B. Position qualification: Ability to maintain facility during incident

C. Position Description
1. Obtain briefing from the support branch director or LSC
2. Participate in logistics meetings
3. Determine requirements for each facility to be established from logistics, OPS, and command
4. Prepare layouts of incident facilities. Post signs, barriers, and security
   a. Communication near Operations
   b. Front tables for Briefing and debriefing of teams
   c. Work space for plans isolated from noise and distractions
   d. Quiet space or separate facility for family members
   e. Separate facility for media
   f. Radio check-in/out near debriefing area

5. Notify unit leaders of facility layout
6. Provide sleeping facilities, security services, facility maintenance services (sanitation, lighting, clean-up)

7. Attempt to establish an area that staff may nap separate from field resources
8. Maintain unit log of donated services and purchased services
9. Demobilize base and camp facilities

X. Ground Support Unit Leader

The Ground Support Unit Leader (transportation) is primarily responsible for transportation of personnel, supplies, food, and equipment, fueling, service, maintenance, repair, and traffic plan.

A. Activation
   1. Activated by Support Branch Director or LSC
   2. Activated whenever need to transport equipment or personnel requires coordination
B. Position qualification: Ability to arrange rides and track vehicles

C. Position Description
   1. Obtain briefing and report to Support Branch Director or LSC
   2. Implement traffic and parking plan developed by planning section. Arrange for and activate fueling, maintenance, and repair of ground resources. Provide transportation services.

UNIT 12: FINANCE/ADMINISTRATION

I. IC Overview of Finance

A. Aggressive initial responses that quickly locate the subject may be more cost effective than a long protracted search with minimal resources

B. A LSC chief who is a volunteer in the local community (such as the local fire chief or rescue squad captain) may be more effective at securing donations than a paid staff person

C. Resources should be looked at for efficiency and not just solely cost per unit time

D. Finances must be looked at as a search factor, but should not be the sole factor

E. The most common position to activate if using paid resources within the Finance/Administration section is the time unit leader. Attempt to find additional roles since this person will be busy only at the beginning and end of the operational period

> **Objectives:**
> ☐ Describe the role and responsibilities of the Finance Section Chief, Time Unit Leader, Procurement Unit Leader, and Compensation Injury Specialist.
> ☐ List situations that would require activation of the elements found within the Finance/Administration Section.
> ☐ Describe methods the IC may use to contain costs.

II. Finance Section Chief

The Finance Section Chief is responsible for all financial accounting, cost analysis of the incident, and for supervising staff.

A. Activation
1. Activated by IC
2. Requirement to provide detailed financial tracking of resources
3. Requirements of agency administrator or base requirements
4. Collection and distribution of money required

B. Position training: Specific FSC training, financial background

C. Position Description
1. Obtain briefing and report to IC
2. Obtain briefings from the agency administrator and other assisting agencies as needed
3. Attend planning meetings to gather information and point out any cost constraints
4. Maintain daily contact with agency'(s) administrative headquarters on financial matters
5. Supervise staff

6. Develop operating plan for Finance
   a. Objectives for staff
   b. Required documentation
   c. Outline reporting and record-keeping requirements

7. Determine if need for personnel time records required on scene
8. Brief personnel on all incident-related business management issues needing attention and follow-up
9. Participate in demobilization planning
10. Ensure all documentation is properly prepared, collected, and completed

III.   Time Unit Leader

The Time Unit Leader is responsible for personnel time recording. This function may be linked to staging area manager on smaller searches.

A. Activation: Activated by FSC, IC, or staging

B. Position qualification: Base qualified, ability to ensure forms correctly filled in

C. Position Description
   1. Obtain briefing from FSC, IC, or staging
   2. Establish work area in close association to sign-in function
   3. Determine which agencies require time unit function, special forms required, special operational requirements
      a. On-scene rest/meal time
      b. Base/field time
      c. On-scene sleeping
      d. Travel time

   4. Ensure time recording documents prepared daily and compliance with time policies met
   5. Submit cost estimate data as required
   6. Provide for security of records
   7. Ensure all records current and complete prior to demobilization
   8. Release reports to agency representative as required

IV. Procurement Unit Leader

The Procurement Unit Leader is responsible for collecting monies/donations during an incident, recording all item contributions, issuing receipts, and administering all financial matters pertaining to vendor contracts. The unit is also responsible for maintaining equipment time records.

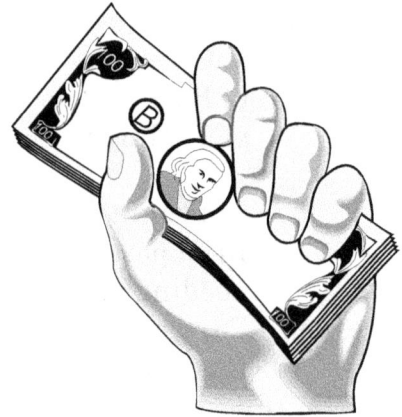

A. Activation: Activated by FSC

B. Position qualification: Position specific training. Ability to handle financial matters.

C. Position Description
1. Report to and obtain briefing from Finance Section Chief
2. Contact appropriate unit leaders on incident needs and any special procedures
3. Develop incident procurement plan
4. Ensure on-scene accounting of all donations generated on-scene
5. Issue receipts and thank-you notes for any equipment or money donated
6. Prepare and sign contracts and land use agreements as needed
7. Draft memorandums of understanding
8. Establish contracts with supply vendors as required. Interpret contracts/agreements and resolve claims or disputes within delegated authority. Finalize all agreements and contracts.
9. Organize and direct equipment time-recording function as needed
10. Complete final processing and send documents for payment. Investigate possible fraud
11. Coordinate cost data in contracts with cost unit leader if required

V. Compensations for Injury Specialist

The Compensation for Injury Specialist is responsible for administering financial matters arising from injuries and deaths occurring on the incident requiring workmans' compensation or a safety report. Many of the specialist's functions are done or partially done in the medical unit.

A. Activation
1. Activated by FSC
2. Activated when accident occurs

B. Position qualifications: Knowledgeable in completing workmans' compensation forms. Knowledgeable in making safety reports.

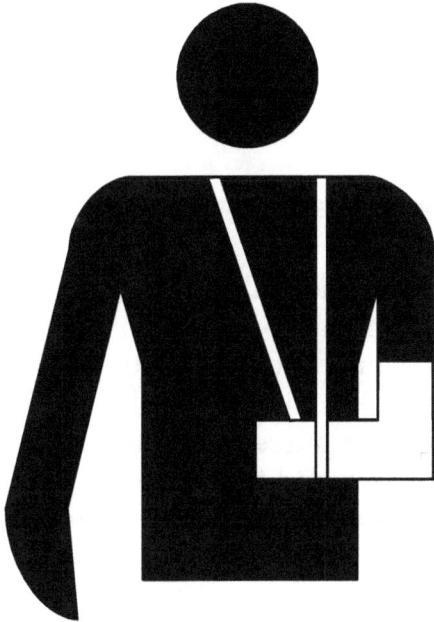

C. Position Description
1. Obtain briefing from and report to FSC
2. Co-locate compensation for injury operations with those of the medical unit when possible
3. Establish contact with incident safety officer, medical unit leader, and agency representatives
4. Ensure that correct agency forms are being used. Provide correct billing forms for transmittal to doctor and/or hospital
5. Keep informed and report on status of hospitalized personnel
6. Obtain all witness statements from safety officer and/or medical unit and review for completeness
7. Provide analysis of injuries and coordinate with safety officer
8. Maintain log of all injuries occurring during incident
9. Arrange for notification of next of kin of seriously injured or deceased persons
10. Coordinate/handle all administrative paperwork on serious injuries or deaths
11. Obtain demobilization plan and insure that necessary follow up action will be handled
12. Coordinate with appropriate agency(s) to assume responsibility for injured personnel in local hospitals after demobilization

UNIT 13: SIMULATIONS

I. Introduction
A. Purpose of simulation
   1. Allow student to practice search management skills
   2. Provide opportunity to work with external influences
   3. Stress created only approximates the stressors of an actual incident

> **Objective:**
> ☐ Demonstrate the ability to effectively manage staff, external influences, and information flow; generate tasks; and efficiently deploy field resources given a search scenario, staff, and field resources.

B. General Paradigm
   1. Class divided in half. Half will serve as staff while the other half serve as field team leaders. Roles will be switched for the second simulation.
   2. Each simulation will run for approximately 1.5 hours
   3. Simulations will operate on accelerated time. A 4-hour field task may only require 10-15 minutes to complete. The team leader will still need to be debriefed and then reassigned.
   4. Clues will be reported to base through the normal channels

II. Simulation Ground Rules

A. Instructors/Staff
   1. When serving in an observer/proctor role the instructors will be appropriately marked. While serving as proctors they may be asked questions and provide assistance. However, they will not make decisions for you
   2. When role playing they will remove the observer/proctor marker. They may don apparel appropriate for the role they are playing. Expect the bizarre and unexpected.

B. Students playing Field Team Leaders
   1. Field Team Leaders are expected to be professional searchers. Conduct should be appropriate for an actual search.
   2. Field Team Leaders must not create problems or situations unless directed by the instructional staff

C. Student Staff
   1. Students will be given basic dispatch information
   2. Students will then be given 5 minutes to organize the management team before arrival on-scene

3. Then let the festivities begin
4. The stress does not equal that of a real search. However, just in case
    a. Be professional at all times
    b. Treat the scenario seriously
    c. No fist-fights
    d. Leave your weapon at the door
    e. Work together
    f. Treat the Field Team Leaders with respect
    g. Try to have fun
    h. Remember this is a learning opportunity

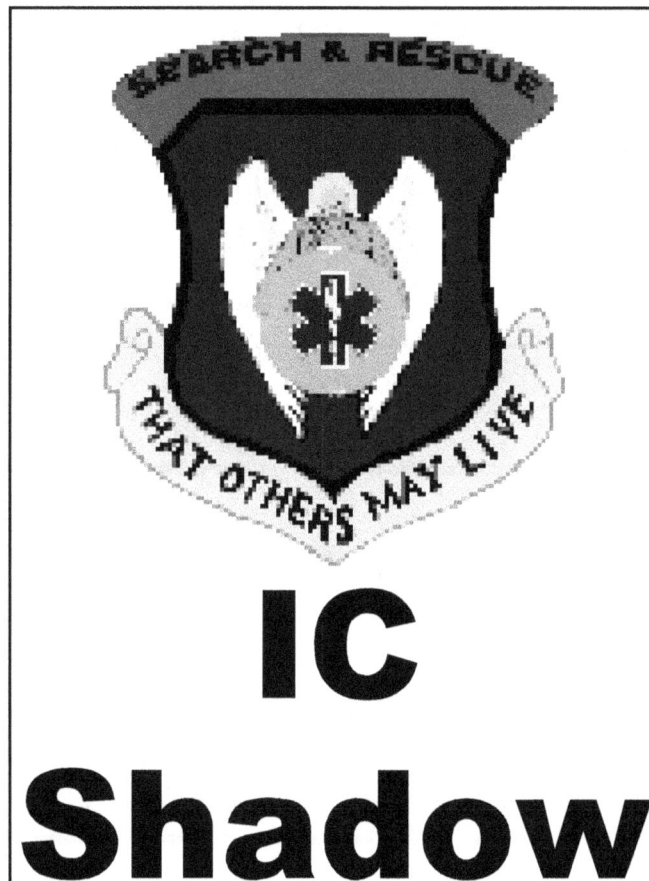

Fig. 13-1 *Name Badge used by Course Staff members used to indicate to students they are in a specific role. When tags are removed they are free to play any role*

UNIT 14: COURSE EVALUATION
Incident Commander for Ground Search and Rescue

Course Number:	
Location:	
Dates:	

Scoring Key

Poor 1	Fair 2	Average 3	Good 4	Excellent 5

Please circle the number that best matches your impression of the particular instructional unit, in each of the three categories. Please take time to explain your response in order to continually improve the course.

1. Course Introduction

Presentation	Materials	Usefulness
1 2 3 4 5	1 2 3 4 5	1 2 3 4 5

Comments:
_____
_____
_____

2. Incident Command System Implementation (Lecture)

Presentation	Materials	Usefulness
1 2 3 4 5	1 2 3 4 5	1 2 3 4 5

Comments:
_____
_____
_____

2. Incident Command System Implementation (Practical)

Presentation	Materials	Usefulness
1 2 3 4 5	1 2 3 4 5	1 2 3 4 5

Comments:
_____
_____
_____

3. Search & Rescue Tactics (Lecture)

Presentation					Materials					Usefulness				
1	2	3	4	5	1	2	3	4	5	1	2	3	4	5

Comments:

3. Search & Rescue Tactics (Practical)

Presentation					Materials					Usefulness				
1	2	3	4	5	1	2	3	4	5	1	2	3	4	5

Comments:

4. Search & Rescue Operations (Lecture)

Presentation					Materials					Usefulness				
1	2	3	4	5	1	2	3	4	5	1	2	3	4	5

Comments:

4. Search & Rescue Operations (Practical)

Presentation					Materials					Usefulness				
1	2	3	4	5	1	2	3	4	5	1	2	3	4	5

Comments:

5. Search & Rescue Resources

Presentation					Materials					Usefulness				
1	2	3	4	5	1	2	3	4	5	1	2	3	4	5

Comments:

6. Legal and Search & Rescue Documents (Lecture)

Presentation					Materials					Usefulness				
1	2	3	4	5	1	2	3	4	5	1	2	3	4	5

Comments:

6. Legal and Search & Rescue Documents (Practical)

Presentation					Materials					Usefulness				
1	2	3	4	5	1	2	3	4	5	1	2	3	4	5

Comments:

7. Aircraft Ground Search & Rescue (Lecture)

Presentation					Materials					Usefulness				
1	2	3	4	5	1	2	3	4	5	1	2	3	4	5

Comments:

7. Aircraft Ground Search & Rescue (Practical)

Presentation					Materials					Usefulness				
1	2	3	4	5	1	2	3	4	5	1	2	3	4	5

Comments:

8. Incident Commander Roles and Responsibilities (Lecture)

Presentation					Materials					Usefulness				
1	2	3	4	5	1	2	3	4	5	1	2	3	4	5

Comments:

8. Incident Commander Roles and Responsibilities (Practical)

Presentation					Materials					Usefulness				
1	2	3	4	5	1	2	3	4	5	1	2	3	4	5

Comments: _____

_____

_____

9. Command Staff (Lecture)

Presentation					Materials					Usefulness				
1	2	3	4	5	1	2	3	4	5	1	2	3	4	5

Comments: _____

_____

_____

9. Command Staff (Practical)

Presentation					Materials					Usefulness				
1	2	3	4	5	1	2	3	4	5	1	2	3	4	5

Comments: _____

_____

_____

10. Search and Rescue Planning Function (Lecture)

Presentation					Materials					Usefulness				
1	2	3	4	5	1	2	3	4	5	1	2	3	4	5

Comments: _____

_____

_____

10. Search and Rescue Planning Function (Practical)

Presentation					Materials					Usefulness				
1	2	3	4	5	1	2	3	4	5	1	2	3	4	5

Comments: _____

_____

_____

11. Logistics (Lecture)

Presentation					Materials					Usefulness				
1	2	3	4	5	1	2	3	4	5	1	2	3	4	5

Comments:

12. Finance/Administration (Lecture)

Presentation					Materials					Usefulness				
1	2	3	4	5	1	2	3	4	5	1	2	3	4	5

Comments:

13. Search Simulation (Practical)

Presentation					Materials					Usefulness				
1	2	3	4	5	1	2	3	4	5	1	2	3	4	5

Comments:

14. Course Evaluation Tool

Format					Usefulness				
1	2	3	4	5	1	2	3	4	5

Comments:

APPENDIX A
Glossary

Aeronautical Fixed Service (AFS): A telecommunication service between specified fixed points provided primarily for the safety of air navigation and for the regular, efficient, and economical operation of air services.

Agency Administrator (AA): Person representing an agency with legal jurisdiction over SAR operations. Synonymous with Legal Responsible Agent.

Agency Representative (AR): An individual assigned to an incident from an assisting or cooperating agency who has been delegated authority to make decisions on all matters affecting the agency's participation at the incident. Agency Representatives report to the Incident Liaison Officer.

Air Route Traffic Control Center (ARTCC): An FAA facility providing air traffic control service principally during enroute flight to aircraft operating on IFR flight plans within controlled airspace.

Air Traffic Control (ATC): A service operated by an appropriate authority to promote the safe, orderly, and expeditious flow of air traffic.

Air-Scent Dog: A search dog that is trained to detect and indicate airborne human scent.

Airborne Warning and Control System (AWACS): USAF Boeing E-3 airborne radar system. Records primary and secondary radar targets. Recorded data is retained for a period not exceeding 10 days.

ALERFA: ICAO equivalent to Alert phase.

Alert Phase: 1)An emergency phase assigned when apprehension exists for the safety of a craft, person, or area because of information that serious difficulties may exist that do not amount to a distress, or because of continued lack of information concerning progress or position. 2) Used to indicate a search is in progress but a specific resource is not requested to respond at this time.

Alert Notice (ALNOT): A message sent by an FSS, ARTCC, or EOC that requests either an extensive communications search for overdue, unreported, missing aircraft/person; or that SAR resources be placed on alert status.

Alert: 1) A status indicating a likelihood of a SAR activation for a SAR resource. 2) Term used to describe the possibility of an air-scent dog detecting human scent. 3)COSPAS-SARSAT report of an apparent distress routed to the search and rescue system

Alert Officer (AO): Person delegated the task of evaluating the situation, determining appropriate response, and initiating dispatch of resources.

Alerting Post: Any facility other than a coast radio station designed to serve as an intermediary between a person reporting an incident or other emergency and an RCC or RSC.

Allocated Resources: Resources dispatched to an incident that have not yet checked-in with the Incident Communications Center. Synonymous with enroute.

Area Command Authority (ACA): Central control person or unit that oversees and allocates resources when multiple incidents occur.

Assigned Resources: Resources checked-in and assigned work tasks on an incident.

Assisting Agency: An agency directly contributing resources to another agency or the incident.

Available Resources: Resources assigned to an incident and available for an assignment within three (3) minutes.

Beacon: Device operating on 121.5, 243, or 406 MHZ intended solely for distress signaling.

Bogus Search: A search that occurs when the subject is not in the search area. Synonymous with "bastard search."

Branch: The organizational level having functional/geographic responsibility for major segments of incident operations. The Branch level is organizationally between section and Division/Group.

Call-Out Qualified (COQ/CQ): A training standard that defines the minimum requirements for personnel to respond to a SAR Mission.

Camp: A geographical site, within the general incident area, separate from the Incident Base, equipped and staffed to provide sleeping, food, water, and sanitary services to incident personnel.

CASIE III: SAR software program to assist in planning and tracking SAR operations.

Check-in: Locations where assigned resources check-in at an incident. The locations are: Incident Command Post, Incident base, Camps, Staging Areas, Helibases, or Division Supervisors.

Clear Text: The use of plain English in radio communications transmissions, NO ten codes, or agency-specific codes are used when using Clear Text.

Command: The act of directing, ordering, and/or controlling resources by virtue of explicit legal, agency, or delegated authority.

Command Post (CP): The location that houses the Incident Commander, Command Staff, and typically Incident Communications.

Command Staff: The Command Staff consists of the Information Officer, Safety Officer, and Liaison Officer, who report directly to the Incident Commander.

Communication units: A facility used to provide the major part of an Incident communication center.

Computer-Aided Search Planning (CASP): A computer search planning system which uses simulation techniques to produce multiple datum points that are displayed as a map of all possible locations.

Continuous Data Recording (CDR): Data extraction terminology associated with FAA ARTSIII and E-ARTS tracking systems.

Cooperating Agency: An agency supplying assistance, other than direct resources, to the incident effort (Red Cross, Telephone company, etc.).

Coordination: The process of systematically analyzing a situation, developing relevant information, and informing appropriate command authority (for its decision) of viable alternatives for selection of the most effective combination of available resources to meet specific objectives. Personnel responsible for coordination may perform command or dispatch functions within limits established by specific agency delegations, procedures, legal authority, etc.

Coordinator of Emergency Services: The person appointed by the governor, as the agency head of the Department of Emergency Services, with the responsibility to coordinate and administer emergency services operations .

COSPAS: Russian segment of the COSPAS-SARSAT satellite system. See SAR Satellite-aided tracking.

Coverage Factor: 1) A measure of search effectiveness or quality, an intermediate calculation when developing Probability of Detection (POD). 2) Ratio of sweep width to track spacing.

Critical Incident Stress Debriefing (CISD): A method of defusing in a systematic fashion search and management resources that have experienced stressful events.

Critical Separation: Spacing of searchers determined by field trial (placement of target item and movement to distance that it just disappears from sight) that determines POD = 50%.

Cumulative Probability of Detection (POD$_{cum}$): The likelihood of finding a target as determined from the mean coverage factor together with the appropriate curve for the number of searches. May be derived from formula or data charts.

Delegation of Authority: Generally written delegation of Authority from the Agency Administrator or Legal Responsible Agent granting the Incident Commander specified control of the agency's resources in order to meet the objectives of the incident.

Despondent: A classification of missing person who is either suicidal or depressed.

DETRESFA: ICAO equivalent to Distress Phase.

Dispatch: The implementation of a command decision to move a resource or resources from one place to another.

Dispatch Center: A facility from which resources are directly assigned to respond to an incident. Dispatch center may also be tasked with tracking resources enroute to and from an incident.

Dispatch Officer (DO): Personnel responsible for the direct operation of the dispatch center.

Dispatch Supervisor (DS): The person responsible for overseeing the operation of the Dispatch Center during SAR Operations.

Distress Phase: An emergency phase assigned when immediate assistance is required by a craft or person because of threat or grave or imminent danger, or because of continued lack of information concerning progress or position after procedures for the Alert phase have been executed.

Division: Divisions are normally established to divide an incident into geographical areas of operation. Divisions are established when the number of resources exceeds the span of control of the Operations Chief.

Dog Team: One trained dog handler and one trained SAR dog. Under most circumstances it will also contain a team walker.

Emergency Locator Transmitter (ELT): Aeronautical distress beacon for alerting and transmitting homing signals.

Emergency Medical Technician (EMT): A level of training established by DOT standards and receiving certification through states. Several different skill levels exist.

Emergency Position-Indicating Radio Beacon (EPIRB): Maritime distress beacon for alerting. It may also transmit homing signals.

Emergency Phase: Any of the phases into which SAR incidents and subsequent SAR missions are classified. See Alert and Distress phase.

Field Team Member (FTM): An individual who has completed certified training to safely and productively participate in SAR operations as a team member on most field tasks.

Field Team Leader (FTL): An individual who has completed training to safely lead, and organize a team of field team members to participate in SAR ops on most search field tasks.

Field Team Signcutter (FTS): An individual who has completed certified training to notice and observe tracks and sign of human passage. Qualified to analyze field clues and determine appropriateness of calling in trained trackers.

Field Operational Unit (FOU): A responding search unit that may deploy resources on field tasks. Similar to a SRU.

Finance/Administration Section Chief (FSC): Part of the General Staff responsible for finances and general administration during an incident.

Flight Information Region (FIR): An airspace of defined dimensions within which flight information service and alerting service are provided.

Flight Information Service (FIS): A service provided for the purpose of giving advice and information useful for the safe and efficient conduct of flights.

Flight Service Station (FSS): An air traffic facility that provides enroute communications and VFR SAR services, assists lost aircraft and aircraft in emergency situations, and originates Notices to Airmen.

Forward-Looking Airborne Radar (FLAR): Any aircraft-mounted radar designed to detect targets on or near the ocean surface by scanning a sector typically centered on the direction of aircraft heading. FLAR may also perform weather-avoidance/navigation in support of aircraft operations

Forward-Looking Infrared Radar (FLIR): An imaging system, mounted on board surface vessels or aircraft, designed to detect thermal energy (heat) emitted by targets and convert into visual display.

Global Positioning Satellite (GPS): A network of satellites that give position to a receiver in latitude and longitude.

General Staff: The group of incident management personnel comprised of Incident Commander, Operations Chief, Planning Chief, Logistics Chief, and Finance/Admin. Chief.

Ground Search and Rescue (GSAR): SAR specifically concerned with the ground elements of searches for missing persons and a missing aircraft search.

Group: Groups are normally established to divide the incident into functional areas of operation. Groups are composed of resources assembled to perform a special function not necessarily within a single geographic division.

Handler: A tracking, trailing, or air-scent dog handler.

Heading: The horizontal direction in which a craft is pointed.

Helibase: the main location, within the general incident area, for parking, fueling, maintenance, and loading of helicopters.

Helispot: A location where a helicopter can take off and land. Often referred to as Landing Zone.

IFF: Outdated recognized term "Identification, Friend or Foe," for radar interrogation of aircraft.

Incident: An occurrence or event, either human-caused or natural phenomena, that requires action by emergency service personnel to prevent or minimize loss of life or damage to property and/or natural resources.

Incident Action Plan (IAP): The plan which lists the various control objectives, safety messages, organizational structure, and other plans needed to run an incident. Similar to the Search Action Plan.

Incident Base: That location at which the primary logistics functions are coordinated and administered. The Incident Command Post may be co-located with the Base. There is only one Base per incident.

Incident Command Post (ICP): The location at which the primary command functions are executed and usually co-located with the incident base.

Incident Command System (ICS): The combination of facilities equipment, personnel, procedures, and communications operating with a common organizational structure, with responsibility for the management of assigned resources to effectively accomplish stated objectives pertaining to an incident.

Incident Commander (IC): The individual responsible for the overall management of all incident operations.

Incident Commander for Ground (ICG): 1) A training standard and qualification system for Incident Commanders trained to handle ground search and rescue incidents. 2) A training course designed to train level three (single agency) Incident Commanders.

Incident Staff (IS): 1) Refers to Command Staff, Operations, and Planning Section Chiefs. 2) Training standard used by some agencies.

Information Officer (IO): A member of the Command Staff responsible for the Public Information function and distributing news and information to those at the incident.

Instrument Flight Rules (IFR): Rules governing the procedures for conducting instrument flight. Also a term used by pilots and controllers to indicate type of flight plan.

Initial Attack: The control efforts taken by resources which are the first to arrive at an incident. During the time period, reflex tasks are often utilized.

Initial Planning Point (IPP): Base point from which median distances are plotted. It may be the same as the Point Last Seen (PLS) or the Last Known Position (LKP).

Jurisdictional Agency: The agency having jurisdiction and responsibility for a specific geographical area. Responsible for providing an AA or RA.

Knot (kt): A unit of speed equal to one nautical mile per hour (1.15 mph).

Last Known Position (LKP): The last known location for the missing subject as determined by verifiable physical evidence such as a discarded object or a footprint. In the case of an aircraft the LKP may be the last reported point or the last observed radar position.

Latitude - Longitude (Lat-Long): A method of fixing a position on the earth. Used by LORAN stations.

Legal Responsible Agent (RA): The official and/or agency having legal responsibility for the emergency response to a SAR incident. In ICS this person is known as the "Agency Administrator."

Liaison Officer: A member of the command staff responsible for interacting with agency representatives from assisting and cooperating agencies.

Local SAR Coordinator: The person responsible for coordination of SAR operations within a given jurisdiction.

Logistics Section Chief (LSC): A member of the General Staff responsible for providing logistical support (Communications, medical, food, supply, facilities, and transportation) to the incident.

Long Range Aid to Navigation (LORAN): An aid to navigation that allows aircraft and boats to determine a latitude and longitude from ground radio stations.

Lost Person Questionnaire (LPQ): Investigative form used to collect information on the lost/missing person.

Management by Objective (MBO): Topdown management so that all involved know and understand the objectives of the operation.

MAYDAY Spoken international distress signal, repeated three times.

MEDEVAC: Air evacuation of person for medical reasons.

Medic: Person with overall responsibility of providing medical care to the sick or injured subject or searcher. Typically certified as an EMT or higher.

MEDICO: Medical consultation. Exchange of medical information and recommended treatment for sick or injured persons where treatment cannot be administered directly by prescribing medical personnel.

Memorandum of Understanding (MOU): A written agreement between parties explaining the expectations of each party.

Message Center: The message center is part of the Incident Communications Center and is co-located or placed adjacent to it. It receives, records, and routes information about resources reporting to the incident, resource status, and administration and tactical traffic.

Mission Coordinator (MC): The person designated by the Legal Responsible Agent or Agency Administrator to coordinate and manage on-scene operations during a specific SAR incident. This person may be called the SAR Mission Coordinator (SMC), On-Scene Commander (OSC), or Incident Commander (IC).

Mission Control Center (MCC): Ground system element of COSPAS-SARSAT that receives data from Local User Terminals, exchanges information with other Mission Control Centers, and distributes alerts and other COSPAS-SARSAT information primarily within its associated service area.

Mobilization Center: An off-incident location at which emergency service personnel and equipment are temporarily located pending assignment, release, or reassignment.

Multiagency Coordination System (MACS): A generalized term which describes the combination of facilities, equipment, personnel, procedures, and communications from various agencies integrated into a common system with responsibility for coordination of assisting agency resources and support to agency emergency operations.

National Track Analysis Program (NTAP): An FAA system for retrieval of computer-stored radar data to locate a missing aircraft's last position.

National SAR Plan (NSP): An interagency agreement providing a national plan for the coordination of SAR services to meet domestic needs and international commitments.

National Interagency Incident Management System (NIIMS): Consists of five major subsystems which collectively provide a total systems approach to all-risk incident management. The subsystems are: The Incident Command System, Training, Qualifications and Certification, Supporting Technologies, and Publications Management.

Nautical Mile (NM): A unit of linear measurement used by ships and aircraft equal to a minute (1/60 of a degree of latitude). A nautical mile equals 1.15 statue miles.

Notification: A level of alert issued to agencies that an incident is ongoing but the likelihood of agency participation is low.

On Scene Commander (OSC): Commander of an SRU assigned to coordinate SAR operations within a specified search area.

Operations Section Chief (OPS): A member of the General Staff responsible for overseeing all operational aspects of an incident and meeting the objectives set in the IAP.

Operational Period: The period of time scheduled for execution of a given set of operation actions as specified in the Incident Action Plan.

Operations Coordination Center (OCC): the primary facility of the Multi-agency Coordination System. Its houses the staff and equipment necessary to perform the MACS functions.

Operations Center: Multi-mission Coast Guard Centers which may function as a RCC.

Orthophoto Maps: Aerial photographs corrected to scale such that geographic measurements may be taken directly from the prints. They may contain graphically emphasized geographic features and may have overlays of such features as: water systems, important facility location, etc.

Other Guy (OG): Informal term for second person on an overhead team responsible for fulfilling all functions not assumed by the Incident Commander.

Out-of-Service Resources: Resources assigned to an incident but unable to respond for mechanical, rest, or personal reasons.

Overhead Personnel: Personnel who are assigned to supervisory position which include Incident Commander, Command Staff, General Staff, Directors, Supervisors, and Unit Leaders.

Overhead Team (OHT): The initial organized and trained team that arrives at a SAR incident and takes operational control of SAR resources.

Paramedic: A level of medical certification used in the civilian world representing the highest level of EMT training.

Planning Meeting: A meeting, held as needed throughout the duration of an incident, to select specific strategies and tactics for incident control operations and for service and support planning.

Plans Section Chief (PSC): A member of the General Staff responsible for planning and documentation of an incident.

Plans and Operations Combined (PLOPS): Functional position when one person fulfills both the Planning and Operations position.

Point Last Seen (PLS): The point the lost person was last seen.

Preventative Search and Rescue (PSAR): Programs developed to decrease the likelihood of persons becoming lost or mitigating risk if the person does become lost.

Probability of Area (POA): The probability that the search object is in the search area.

Probability of Clue (POC): The probability that a clue relates to the missing subject.

Probability of Detection (POD): The probability that the search object will be detected provided it is in the area searched. Measures search results.

Probability of Success (POS): The probability the search object is in the search area and that it will be located. Measure of search effectiveness.

Public Information Officer (PIO): A member of the Command Staff responsible for media contacts and control.

Public Service Announcement (PSA): A method of PSAR to provide the public with information to mitigate possible SAR incidents.

Radarfind (RADF): Computer program designed to rapidly locate available recorded radar data to assist in the location of missing/downed aircraft.

Radio Cache: A cache may consist of a number of portable radios, a base station and in some cases a repeater stored in a predetermined location for dispatch to incidents.

Reflex Task: Initial tasks sent out during the opening of a search that follow standard scenario-based patterns and are sent into high POA sectors.

Relative Urgency Response Factors (RURF): Computer software that calculates the urgency of a missing person search based upon several factors and determines a level of initial urgency.

Relevance of Clue (ROC): Value in rank order of located clues.

Resources: All personnel, services, and major items of equipment available, or potentially available, for assignment to incident tasks on which a status is maintained.

Rescue Coordination Center (RCC): A unit responsible for promoting efficient organization of SAR services and coordinating conduct of SAR operations within a Search and Rescue Region (SRR).

Rescue Unit (RU): A unit composed of trained personnel and provided with equipment suitable for expeditious conduct of rescue operations.

Rescue Specialist (RS): An individual who has completed standardized training and is qualified to carry out rescue procedures.

Resource Mixing: Application of different resource types to a search sector usually at different times to balance their strengths and weakness and raise the real PODcum.

Rest of World (ROW): The probability that the subject or target is not in the search area.

RESTAT: An acronym for the Resources Unit- a unit within the planning section responsible for tracking resources assigned to an incident.

Rural SAR: Search and Rescue operations conducted in a farming, woodland, or sparsely to moderately populated area.

Safety Officer (SO): A member of the Command Staff responsible for the overall safety of the Incident. They normally work within the proper chain of command but have emergency power to stop unsafe practices.

SAR Coordinator (SC): The agency or official responsible for the SAR organization and coordination of SAR operations in a given area or region.

SAR Emergency: Any SAR incident, whether related to any other type of incident or not, that requires the utilization of resources to resolve, due to the threat or potential threat to human life.

SAR Incident: Any situation requiring notification and alerting of the SAR system and which may require SAR mission(s).

SAR Mission Coordinator (SMC): The person designated by the legal responsible agent to coordinate and manage on-scene operations during a specific SAR incident. This person may be called the Mission Coordinator (MC), On-Scene Commander (OSC), or the Incident Commander.

Scenario Analysis: A modification to the Mattson consensus process to adjust for multiple scenarios resulting in different areas of probability.

Search Action Plan (SAP): 1) Message, normally developed by the SMC, passing instructions to SRUs and agencies participating in a SAR mission. 2) Similar to an Incident Action Plan, listing the various objectives, organizational structure, and plans need to run an incident.

Search and Rescue System: An arrangement of components activated as needed to efficiently and effectively aid persons or property in actual or potential distress.

Search and Rescue (SAR): The use of available resources to assist persons and property in potential or actual distress.

Search and Rescue Region (SRR): A defined area with an RCC, within which SAR services are provided.

Search and Rescue Satellite Aided Tracking (SARSAT): Part of an international system of satellites and ground network stations for distress alerting, positioning, and use of emergency beacons. Combined with the Russian segment, COSPAS, to form the COSPAS-SARSAT system.

Search Area: Area assigned by competent authority to be searched.

Search Master (SM): Canadian term for an individual appointed to coordinate and direct a specific SAR operation.

Search Pig: A pig that located people by scenting on truffles the missing subject carries at all times. Handlers found at "watering holes."

Search Pattern: A track line or tactical procedure assigned to a search strike force, task force, or SRU for searching a specified sector.

Search and Rescue Unit (SRU): A specialized unit organized for SAR. Synonym for Field Operational Unit (FOU).

Section: That organizational level having functional responsibility for primary segments of the incident such as: Operations, Planning, Logistics, Finance/Administration. The section is organizationally between branch and Incident Commander.

Sector: 1) A search area in which a SRU may be deployed to accomplish a single sortie or task in a 4-8 hour timespan. 2) A defined area in an RSC, within which SAR services are provided.

Sector Ladder: Simplified method of prioritizing search sectors, deploying resources, and determining when to re-search an sector.

Sector X: Sector that represents the area outside of the search area.

Segment: A search area that my contain multiple sectors. Often used for planning purposes.

Side-Looking Airborne Radar (SLAR): Aircraft-mounted radar designed to detect targets on or near the ocean surface by transmitting signals perpendicular to the aircraft flight track. Signal coverage of an area is achieved by aircraft motion alone, without antenna rotation.

SITSTAT: An acronym for the Situation Unit. A unit within the planning section responsible for keeping tract of incident events.

Sortie: Individual movement of a resource in rendering assistance. Often referred to as a search task.

Sound Sweep: A search technique of increasing missing person detection of responsive subjects by systematically signaling or shouting.

Span-of-Control: The supervisory ratio of from three to seven individuals which someone may effectively supervise, with five being the established general rule of thumb.

Staging Area: A temporary on-incident location where incident personnel and equipment are assigned on a (3) minute available status.

Strike Team: Specified combinations of the same kind and type of resources, with common communications and a leader.

Task Assignment Form (TAF): A form used to dispatch, track, document, and debrief tasks or sorties.

Task Force: Any combination of resources with common communications and a leader.

Task Package: Package that contains TAF, task map, subject information sheet, and any other information given to a field team to complete a field task.

Technical Specialist: Personnel with special skills who are activated only when needed.

Topographical Map (Topo): A map which depicts changes in elevation, land features, hydrological features, and man-made features. The 7.5-minute and 15-minute scale are commonly used in missing person incidents.

Track Trap: An area where tracks can be easily detected.

Tracker: An individual who can detect and follow signs of human passage. Sometimes also referred to as a "mantracker".

Tracking Dog: A search dog that will follow the ground scent of a person who has passed through an area in which the dog is searching.

Trailing Dog: A search dog that will follow the scent trail of a specific individual, after the dog has been allowed to smell an article or object that has been in contact with that individual. The trailing dog is scent discriminating.

True Air Speed (TAS): The speed an aircraft is making through the air.

Unified Command: A method for all agencies or individuals who have jurisdictional responsibility, and in some cases those who have functional responsibility at the incident, to contribute to determining overall objectives for the incident, and selection of a strategy to achieve the objectives.

Unit: The organizational element having functional responsibility for a specific incident planning, logistics, or finance activity.

Universal Time Coordinated (UTC): International term for time at the prime meridian (once called Greenwich Mean Time.)

Urban/Suburban SAR: Search and Rescue operations conducted in moderately to heavily populated areas given primarily to business and residential developments.

Urban Search and Rescue Task Force (USAR TF): A FEMA task force organized for search and rescue in collapsed buildings.

Wilderness Emergency Medical Technician (WEMT): A level of standardized training providing EMTs with specialized wilderness training and protocols.

Wilderness SAR: SAR operations conducted in an area generally remote and uninhabited and often inaccessible by road.

APPENDIX B
List of Abbreviations

A

A	Search Area
AA	Agency Administrator
A/C	Aircraft
AA	Agency Administrator
ACA	Area Command Authority
ACV	Air Cushion Vehicle
ADCOM	Air Defense Command
ADF	Automatic Direction Finding
AECC	Aeromedical Evacuation Control Center
AFB	Air Force Base
AFRCC	Air Force Rescue Coordination Center
AGIL	Airborne General Illumination Lightship
ALERFA	Alert Phase (ICAO)
ALNOT	Alert Notice
AO	Alert Officer
AR	Agency Representative
ARES	Amateur Radio Emergency Services
ARC	American Red Cross
ARS	Air Rescue Service
ARTCC	Air Route Traffic Control Center
ASRC	Appalachian Search & Rescue Conference
ASTM	ASTM Inc.
ATC	Air Traffic Control
ATCC	Air Traffic Control Center
ATS	Air Traffic Service
ATV	All Terrain Vehicle
AWACS	Airborne Warning and Control System

B

B/D	Briefing/Debriefing Function
BQ	Base Qualified
BO	Bike Operator
BRO	Base Radio Operator
BSA	Boy Scouts of America

C

CAP	Civil Air Patrol
CASARA	Canadian Air Search and Rescue Association
CASIE III	Computer Aided Search Information Exchange Ver III.
CASP	Computer Aided Search Plan
CDR	Continuous Data Recording
CG	Channel Guard
CGAS	Coast Guard Air Station
CGAUX	Coat Guard Auxiliary
CISD	Critical Incident Stress Debriefing
COMCEN	Communications Center
COQ/CQ	Call-Out Qualified
COSPAS	Cosmicheskaya Sistyema Poiska Avariynych Sudov
CP	Command Post
CPR	Cardiopulmonary Resuscitation
CSP	Commence Search Point
CUL	Communications Unit Leader

D

DAN	Diver's Alert Network
DES	Department of Emergency Services
DETRESFA	Distress Phase (ICAO)
DF	Direction Finding
DH	Dog Handler
DMAT	Disaster Medical Assistance Team
DME	Distance Measuring Equipment
DO	Dispatch Officer
DOA	Dead on Arrival
DOD	Department of Defense
DOI	Department of Interior
DOT	Department of Transportation

E

ECC	Emergency Communications Center
ECO	Emergency Care Order

ELT	Emergency Locator Transmission
EMS	Emergency Medical Service
EMT	Emergency Medical Technician
EOC	Emergency Operations Center
EOP	Emergency Operations Plan
EPIRB	Emergency Position-Indicating Radio Beacon
ERI	Emergency Response Institute
ESAR	Explorer Search and Rescue
ETA	Estimated Time of Arrival
ETD	Estimated Time of Departure
ETI	Estimated Time of Intercept
evac	Evacuation

F

FAA	Federal Aviation Agency
FAR	Federal Aviation Regulation
FBI	Federal Bureau of Investigations
FCC	Federal Communications Commission
FEMA	Federal Emergency Management Agency
F^f	Fatigue Correction Factor
FIS	Flight Information Service
FLAR	Forward - Looking Airborne Radar
FLIR	Forward Looking Infrared
FLO	Family Liaison Officer
FOG	Field Operations Guide
FOGSAR	*Field Operations Guide for SAR*
FOR	Field Operational Resource
FOU	Field Operational Unit
FSC	Finance Section Chief
FSS	Flight Service Station
FTL	Field Team Leader
FTM	Field Team Member
FTS	Field Team Signcutter

G

GPS	Global Positioning Satellite
GS	General Staff
GSA	Girl Scouts of America
GSAR	Ground Search and Rescue

H

HAT	Hug-A-Tree
HF	High Frequency
HQ	Headquarters

I

IAP	Incident Action Plan
IC	Incident Commander
ICAO	International Civil Aviation Organization
ICG	Incident Commander for Ground
ICGSAR	*Incident Commander for Ground Search and Rescue*
ICP	Incident Command Post
ICS	Incident Command System
ICSAR	Interagency Committee on Search and Rescue
IFF	Identification, Friend or Foe
IFR	Instrument Flight Rules
INMARSAT	International Maritime Satellite
IO	Information Officer
IP	Initial Position
IPP	Initial Planning Point
IR	Infrared
IS	Incident Staff

J

JRCC	Joint Rescue Coordination Center

K

kt	Knot (Nautical Miles per Hour)

L

LKP	Last Known Position
LORAN	Long-Range Aid to Navigation
LPQ	Lost Person Questionnaire
LSC	Logistics Section Chief
LZ	Landing Zone

M

MACS	Multi-agency Coordination System
MAST	Military Assistance to Safety and Traffic
MBO	Management by Objective
MC	Mission Coordinator

MEDEVAC	Medical Evacuation		**OG**	Other Guy
MEDICO	International word meaning a radio medical situation		**OHT**	Overhead Team
			OPCEN	Coast Guard Operations Center
MLPI	*Managing the Lost Person Incident*		**OPS**	Operations Section Chief
			OSC	On-Scene Commander
MOA	Military Operating Area		**OSE**	On-Scene Endurance
MOTV	Modified Offset and Track Variable		**OSHA**	Occupational Safety and Health Agency
MOU	Memorandum of Understanding			
MPH	Miles per Hour			**P**
MPQ	Missing Person Questionnaire		**PA**	Physician's Assistant
MRA	Mountain Rescue Association		**PADI**	Professional Association of Diving Instructors
MRE	Meals Ready to Eat			
MSF	*Managing the Search Function*		P^c	Cumulative Probability of Detection
MSO	Managing Search Operations			
MVFR	Mountainous VFR		Pd_n	Probability Density
			PFD	Personal Floatation Device
	N		**PIO**	Public Information Officer
NAS	Naval Air Station		**PIW**	Person in Water
NASAR	National Association for Search and Rescue		**PJ**	Para-rescueman (USAF)
			PL	Private Line
NAUI	National Association of Underwater Instructors		**PLB**	Personal Locator Beacon
			PLOPS	Plans and Operations Combined
NAVSAT	Navigation Satellite		**PLS**	Point Last Seen
NCIC	National Crime Information Center		**PMA**	Positive Mental Attitude
			POA	Probability of Area
NIIMS	National Interagency Incident Management System		**POB**	Persons on Board
			POC	Probability of Clue (relating to the subject)
NM	Nautical Mile			
NOAA	National Oceanic and Atmospheric Administration		**POD**	Probability of Detection
			POD_{cum}	Cumulative Probability of Detection
NOTAM	Notice to Airmen			
NPS	National Park Service		**POS**	Probability of Success
NSM	National SAR Manual		**POV**	Personally Owned Vehicle
NSP	National SAR Plan		**PRU**	Pararescue Unit
NSP	National Ski Patrol		**PSA**	Public Service announcement
NSS	National SAR Secretariat (CA)		**PSAR**	Preventative Search and Rescue
NTAP	National Track Analysis Plan		**PSC**	Plans Section Chief
NTSB	National Transportation and Safety Board		**PSO**	Practical Search Operations
NWS	National Weather Service			**Q**
			QRT	Quick Response Team
	O			
OCC	Operations Coordination Center			**R**
ODH	Operational Dog Handler		**R**	Search Radius
OES	Office of Emergency Services		**RA**	Responsible Agent
OFDA	Office of Foreign Disaster Assistance		**RACES**	Radio Amateur Civil Emergency Services

RADF	Radarfind
RCC	Rescue Coordination Center
RDF	Radio Direction Finder
RESTAT	Resources Unit
RO	Radio Operator
ROC	Relevance of Clue
ROW	Rest of the World
RS	Rescue Specialist
RU	Rescue Unit
RURF	Relative Urgency Response Factors

S

SAP	Search Action Plan
SAR	Search and Rescue
SARDO	Search and Rescue Duty Officer
SARMIS	Search and Rescue Management Information System
SARSAT	Search and Rescue Satellite-Aided Tracking
SARTA	Search and Rescue Training Associates
SC	SAR Coordinator
SC	Signcutter
SCO	State Coordination Officer
SEAL	Navy Sea-Air-Land Unit
SIS	Subject Information Sheet
SITREP	Situation Report
SITSTAT	Situation Unit
SLAR	Side-looking Airborne Radar
SM	Search Master (Canadian)
SMC	Search Mission Coordination
SOP	Standard Operating Procedures
SRR	Search and Rescue Region
SRS	Search and Rescue Sector
SRU	Search and Rescue Unit
ST	Strike Team
STOP	Stop, Think, Observe, Plan
SU	Search Unit
SUV	Sports Utility Vehicle

T

TAF	Task Assignment Form
TAS	True Air Speed
TCA	Terminal Control Area
TDO	Temporary Detaining Order
TFR	Temporary Flight Restrictions
Tk	Tracker

Topo	Topographic Map

U

UHF	Ultra-High Frequency
UMS	Uniform Map System
USC	United States Code
USCG	United States Coast Guard
USAF	United States Air Force
USFS	United States Forest Service
USMC	United States Marine Corps
USN	United States Navy
UTC	Universal Time Coordinated
UTM	Universal Transverse Mercator Grid

V

VFR	Visual Flight Rules
VIP	Very Important Person
VOR	Very High Frequency Omnidirectional Range Station
VORTAC	VHF Omnidirectional Range station/Tactical Air Navigation
VSP	Virginia State Police

W

W	Wind
WEMT	Wilderness Emergency Medical Technician
Wx	Weather

X

X	Sector X for Rest of World (ROW)

Z

Z	Zulu time

APPENDIX C

Standard ICS Search Mission Symbols and Colors
Suggested for placement on base maps.

Color	Symbol	Description	Notes
Black		Sector/planning boundaries	
Black	_/_/_/_/_/_/	Travel barriers	Cliffs, etc.
Black	======== →,bd, etc.	Modifications/updates to map	
Black	[I] [II] (A) (B)	Branches Divisions	Consider naming divisions North, South, East, etc
Red	PLS 9 Jan 1820 ⊗	Point Last Seen or Last known position	Consider adding direction of travel
Red	X	Hazard	Write description
Blue	◣	Incident Command Post	
Blue	Ⓑ	Incident Base	Often same as CP
Blue	Ⓢ	Staging area	Often same as CP
Blue	● H-1	Helispot (location and #)	(LZ)
Blue	Ⓗ	Helibase	
Blue	Ⓡ	Repeater/mobile relay	Add task number if staffed
Blue	Ⓣ	Telephone	
Red	✛	First-Aid Station	
Blue	------------	Theoretical Search Area	Planning Map
Blue	___95%___	Statistical Search Area	Planning Map

Suggested for placement on overlays

Color	Symbol	Description	Notes
Black	→W/5 1800 9 Jan	Wind speed and direction	
Black	[I] [II] (A) (B)	Branches Divisions	Consider naming divisions North, South, East, etc
Red	#1 ©	Clue with clue tracking #	may add time, date
Red	#2 **A** →	Air-scent dog alert (arrow points with dogs nose)	may add time date
Resource specific	/////////////// ///////////////	First coverage of sector	Consider density of lines to reflect POD
Resource specific	\\\\\\\\\\ \\\\\\\\\	Second coverage of sector	Consider density of lines to reflect POD
Resource specific	| | | | | | | | | | | | | | | |	Third coverage of sector	Consider density of lines to reflect POD
Resource specific	======== ========	Fourth coverage of sector	Consider density of lines to reflect POD
Resource specific	⌀⁹⁻²ᐟ 10:20 →	Hasty Task (hash mark indicates current position with time reported). Task # links to task log and TAF.	
Resource specific		Sector Task. Task # links to task log and TAF.	

Resource specific colors*:
Green: Field teams
Brown: Air-scent dog team
Yellow: Tracking/trailing dog
Purple: Horses/helicopters
Orange: Special resources (containment, trackers, 4WD, surveillance, etc.)
Blue: Water teams (SCUBA, air-scent water dog, etc.)

*Since markers get lost or may not be available, the most important factor is to place a color key at the bottom of all maps.

Appendix D
Forms Used and Sources

Form name	Form number	Form source
Incident Briefing	ICS form 201	Oklahoma University
Incident Objectives	ICS form 202	Oklahoma University
Organization Assignment List	ICS form 203	Oklahoma University
Division Assignment List	ICS form 204	Oklahoma University
Communications Plan	ICS form 205	Oklahoma University
Medical Plan	ICS form 206	Oklahoma University
Subject Information Sheet	dbS-301	dbS Productions
Media Plan	dbS-302	dbS Productions
Family Plan	dbS-303	dbS Productions
Demobilization Plan	dbS-304	dbS Productions
Rescue/Evacuation Plan	dbS -305	dbS Productions
Medical Plan Part B	dbS-306	dbS Productions
SAR Incident Organization Chart	dbS-307	dbS Productions
Follow-Up Log	dbS-308	dbS Productions
Incident Organizational Chart	ICS form 207	Oklahoma University
Unit Log	ICS form 214	Oklahoma University
Task Assignment Form (TAF)		dbS Productions
Lost Person Questionnaire (LPQ)		VA SAR Council
Clue Log		dbS Productions
Task Log		VA DES
SAR Unit Personnel Sign-In		VA DES
SAR Volunteer Sign-In		VA DES
Vehicle Register		VA DES
Mission Operational Period Report		VA DES
After Action Report		VA DES
Communications Log		ASRC
Communications Team Tracking Log		dbS Productions
Communications Equipment log		ASRC
Personnel On Scene		ASRC

Appendix E
Common Form Samples*

* Forms have been reformatted to fit the margins of this publication. Non-ICS forms are copyright protected and may be obtained directly from dbS Productions.

INCIDENT OBJECTIVES	1. INCIDENT NAME	2. DATE PREPARED	3. TIME PREPARED

4.   OPERATIONAL PERIOD (DATE/TIME)

5.   GENERAL OBJECTIVES FOR THE INCIDENT (INCLUDE ALTERNATIVES)

6.   WEATHER FORECAST FOR OPERATIONAL PERIOD

7.   GENERAL/SAFETY MESSAGE

8.   ATTACHMENTS (✓ IF ATTACHED)

☐ ORGANIZATION LIST (ICS-203)   ☐ MEDICAL PLAN(ICS-206)   ☐ <u>Safety Message</u>

☐ ASSIGNMENT LISTS (ICS-204)   ☐ INCIDENT MAP   ☐ _____

☐ COMMUNICATIONS PLAN (ICS-205)   ☐ TRAFFIC PLAN   ☐ _____

ICS-202 04-94	9. PREPARED BY (PLANNING SECTION CHIEF)	10. APPROVED BY (INCIDENT COMMANDER)

ORGANIZATION ASSIGNMENT LIST

1. INCIDENT NAME	2. DATE PREPARED	3. TIME PREPARED

POSITION	NAME	4. OPERATIONAL PERIOD (DATE/TIME)

5. INCIDENT COMMANDER AND STAFF

Position	Name
INCIDENT COMMANDER	
DEPUTY	
SAFETY OFFICER	
INFORMATION OFFICER	
LIAISON OFFICER	

6. AGENCY REPRESENTATIVE

7. PLANNING SECTION

Position	Name
CHIEF	
DEPUTY	
RESOURCES UNIT	
SITUATION UNIT	
DOCUMENTATION UNIT	
DEMOBILIZATION UNIT	

TECHNICAL SPECIALISTS

8. LOGISTICS SECTION

Position	Name
CHIEF	
DEPUTY	

a. SUPPORT BRANCH

Position	Name
DIRECTOR	
SUPPLY UNIT	
FACILITIES UNIT	
GROUND SUPPORT UNIT	

b. SERVICE BRANCH

Position	Name
DIRECTOR	
COMMUNICATIONS UNIT	
MEDICAL UNIT	

9. OPERATIONS SECTION

Position	Name
CHIEF	
DEPUTY	

a. BRANCH I-DIVISIONS/GROUP

Position		Name
BRANCH DIRECTOR		
DEPUTY		
DIVISION/GROUP		
DIVISION/GROUP		
DIVISION/GROUP		
DIVISION/GROUP		
DIVISION/GROUP		

b. BRANCH II-DIVISIONS/GROUPS

Position		Name
BRANCH DIRECTOR		
DEPUTY		
DIVISION/GROUP		
DIVISION/GROUP		
DIVISION/GROUP		
DIVISION/GROUP		
DIVISION/GROUP		

c. BRANCH III-DIVISIONS/GROUPS

Position		Name
BRANCH DIRECTOR		
DEPUTY		
DIVISION/GROUP		
DIVISION/GROUP		
DIVISION/GROUP		

d. AIR OPERATIONS BRANCH

Position	Name
AIR OPERATIONS BR. DIR.	
HELICOPTER COORDINATOR	
FIXED WING COORDINATOR	
HELICOPTER COORDINATOR	

10. FINANCE SECTION

Position	Name
CHIEF	
DEPUTY	
TIME UNIT	
PROCUREMENT UNIT	
COMPENSATION/CLAIMS UNIT	

ASSIGNMENT LIST	3. INCIDENT NAME	DATE PREPARED	TIME PREPARED

1. BRANCH	2. DIVISION/GROUP	4. OPERATIONAL PERIOD DATE_____ TIME_____

5. OPERATIONS PERSONNEL

OPERATIONS CHIEF_____   DIVISION/GROUP SUPERVISOR_____

BRANCH DIRECTOR_____   _____   AIR OPERATIONS DIRECTOR_____

6. RESOURCES ASSIGNED THIS PERIOD

STRIKE TEAM/TASK FORCE RESOURCE DESIGNATOR	LEADER	NUMBER PERSONS	TRANS. NEEDED	DROP OFF PT./TIME	PICK UP PT./TIME

7. TACTICAL OPERATIONS
--

8. SPECIAL INSTRUCTIONS

9. DIVISION/GROUP COMMUNICATIONS SUMMARY

FUNCTION		FREQ.	SYSTEM	CHAN	FUNCTION		FREQ.	SYSTEM	CHAN.
COMMAND	LOCAL				SUPPORT	LOCAL	----	----	---
	REPEAT					REPEAT	----	----	---
DIV./GROUP TACTICAL		----	----	----	GROUND TO AIR			----	---

ICS-204 04-94	PREPARED BY (RESOURCES UNIT LEADER	APPROVED BY (PLANNING SECTION CHIEF)

INCIDENT RADIO COMMUNICATIONS PLAN	1. INCIDENT NAME	2. DATE/TIME PREPARED	3. OPERATIONAL PERIOD DATE/

4. BASIC RADIO CHANNEL UTILIZATION					
SYSTEM/CACHE	CHANNEL	FUNCTION	FREQUENCY	ASSIGNMENT	REMARKS

ICS-205
04-94

5. PREPARED BY (COMMUNICATIONS UNIT)

MEDICAL PLAN	1. INCIDENT NAME	2. DATE PREPARED AND TIME PREPARED	4. OPERATIONAL PERIOD

5. INCIDENT MEDICAL AID STATIONS			
MEDICAL AID STATIONS	LOCATION	PARAMEDICS	
		YES	NO

6. TRANSPORTATION

A. AMBULANCE SERVICES

NAME	ADDRESS		PARAMEDICS	
			YES	NO

B. INCIDENT AMBULANCES

NAME	LOCATION	PARAMEDICS	
		YES	NO
None	N/A		

7. HOSPITALS

NAME	ADDRESS (Numbers are Loran coordinates)	TRAVEL TIME		PHONE	HELIPAD		BURN CENTER	
		AIR	GROUND		YES	NO	YES	NO

8. MEDICAL EMERGENCY PROCEDURES

ICS-206 04-94	9. PREPARED BY (MEDICAL UNIT LEADER)	10. REVIEWED BY (SAFETY OFFICER)

Medical Plan Part B	1. Incident name	2. Date prepared	3. Time prepared	4. Operational period

5. Incident Medics

Name	Medical Training	Organization	Location/Assignment

6. Medical Equipment (attach additional sheets if required) ❑additional sheets

Equipment	Current location	Equipment	Current location

7. Transportation to Medical Site

Medical Plan-B dbS 305 1/96	8. Prepared by (Medical unit leader)	9. Reviewed by (Safety Officer)

Rescue/Evacuation Plan	1. Incident Name	2. Date Prepared	3. Time prepared	4. Operational Period

5. Possible Scenarios

❏ High Angle Rescue ❏ Hi-line ❏ Cave ❏ Multi-casualty #_____
❏ Semi-technical ❏ Extrication ❏ Swiftwater ❏ Other_____

6. Rescue Specialist Qualified	7. Location/Assignment	8. Availability

9. Rescue/Evacuation Equipment (attach additional sheets if required) ❏ additional sheets

Equipment	Location	Owner	Equipment	Location	Owner

10. Transportation to Rescue Site.

11. Rescue/Evacuation Procedures

Rescue/Evac plandbS 305 4/96	12. Prepared by (medical unit leader)

Media Plan	1. Incident name	2. Date prepared	3. Time prepared	4. Operational Period

5. Media Contacts ❏ additional sheets

6. Media Organizations	7. Contact person	8. Contact number	9. Deadlines

10. Regular Briefings

Times	Participants	Location(s)	Preparations

11. Media Procedures for Subject(s) found alive and well

12. Media Procedures for Subject(s) found injured

13. Media Procedures for Subject(s) found deceased

14. Additional Procedures

Media plan dbS 302 1/96	15. Prepared by (IO)

Family Plan	1. Incident Name	2. Date Prepared	3. Time Prepared	4. Operational Period

5. Family Members ❏ additional sheets

Name	Relationship	Contact Number	or Location

6. Procedures if subject found alive and well

❏ Primary Contact(s)_____

❏ Notifier _____

7. Procedures if subject(s) found injured

❏ Primary Contact(s)_____

❏ Notifier _____

8. Procedures if subject(s) found dead

❏ Primary Contact(s) _____

❏ Notifier _____

9. Additional Procedures

Family Plan dbS 303 1/9	10. Prepared by (Family Liaison Officer)

Demobilization Plan	1. Incident Name	2. Date Prepared	3. Time Prepared	4. Operational Period

5. Contact Organizations to Stop/Adjust Incoming Resources ❏ additional sheets

Organization	Resource Type	Contact #	Contact Name	Stop time

6. Release Priorities (organizations, logistics, etc)

7. Equipment to Return ❏ additional sheets

Equipment/item	Owner	Contact #	Release Priority

8. Personnel Release Procedures

Field Resource Removal	Field Facilities Support	Incident Debriefing Time:	Base Breakdown	Safety Check

9. Additional Procedures

Demobilization dbS 304 4/96	10. Prepared by (Demobilization Unit Leader

1. Resource Type: —————
2. Planning #: —————
3. Priority: —————

4. Task Completed ❏
5. Task Partially Finished❏
6. URGENT Follow-Up ! ❏

TASK ASSIGNMENT FORM	7. Task Number	8. Team Identifier	9. Resource Type	10. Task Map(s)
11. Branch	12. Division/Group		13. Incident Name	

A S S I G N M E N T

14. Task Instructions

16. Previous Search Efforts in Area

17. Transportation   18. Equipment Requirements

15. Briefing Checklist:
❏ Expected Time frame
❏ Target POD subject
❏ Target POD clues
❏ Teams nearby
❏ Applicable clues
❏ Terrain/Hazards
❏ Weather, Safety Issues
❏ Press, Family Plans
❏ Subject Information
❏ Rescue/Find Plans
❏ Others

19. Role	Name	Agency	Role	Name	Agency
1. FTL			8.		
2.			9.		
3.			10.		
4.			11.		
5.			12.		
6.			13.		
7.			14.		

20. Team Call Sign   Freq.

21. Base Call Sign   Freq.

22. Pertinent Phone Numbers
Base:

23. Instructions
Check in every _____ on the _____ hour.

24. Function	Freq.	Comments	Function	Freq.	Comments
Tactical I					
Tactical II					

25. Notes/Safety Message:

SAR TAF 5	26. Prepared by:	27. Briefed by:	Time out:

TEAM DEBRIEFING	1. Date/Time Prepared	2. Time Task Completed	3. Debriefer Name	4. Task #

W X	5. Wind Direction	6. Wind Speed	7. Wind Variability	8. Temp	9. Cloud Cover	10. Precipitation

11. Resource Type	12. Search Technique	13. Team Performance *explain:* *in block 15 or 17*) ❑ Adequate Equipment ❑ Able to search again ❑ Adequate Composition/Morale ❑ Problem free task

TASK RESULTS

14. Describe Area Actually Searched and How

❑ Task partially finished
❑ Task completed as assigned

15. Describe Areas Not Searched/ Lower PODs

16. Describe Clues, Tracks, Alerts, or Scent Trails (*record on clue log and map*)

Follow-Up Urgency: ❑ High ❑ Medium ❑ Low

17. Describe Follow-Up Recommendations/FTL Comments (*record on F-U log and map*)

Follow-Up Urgency: ❑ High ❑ Medium ❑ Low

18. Hazards/Terrain Noted	19. POD Summary FTL ——— POD Margin of Error	20. POD Summary (Debriefer)
	Responsive Subject	_____
	Unresponsive Subject	_____
	Clues	_____

DEBRIEFER'S SECTION

21. Follow-Up Suggestions

Follow-Up Urgency: ❑ High ❑ Medium ❑ Low

22. Task Summary ❑ Nothing Significant Found ❑ Needs Review_____ ❑ Needs Urgent Review_____	23. Routing for Review Initials Initials ❑ Operations Chief ❑ Documentation _____ ❑ Plans ❑ IC _____ ❑ Investigation ❑ _____ _____

Communication's Team Tracking Log			1. Incident Name		2. Date Begun		3. Operational Period	
Team Designator	Task Number	Freq.	Time Start Task	Last Coord.	Time Last Check-In	Est. % Task Completed	Time Back

Notes:

Clue Tracking Log	1. Incident name		2. Date Begun	3. Operational Period		4. Page _____ of _____ Pages

Clue #	Task #	Time	Map Coord.	Clue Description	Action Taken	Initial

Follow-Up Tracking Log	1. Incident name		2. Date Begun	3. Operational Period	4. Page _____ of _____ Pages	

F/U #	Task #	Time	Map Coord.	Follow-up Required Description	Follow-up Action Taken	Initial

Task Log	1. Incident name		2. Date Begun	3. Operational Period		4. Page _____ of _____ Pages

Task #	Team Identifier	Team Leader	Team Type	# on Team	Task Description	Time Out	Time In	Total Hours

Appendix F

Training Courses

Incident Commander for Ground Search & Rescue   24 hours
Alzheimer's Disease, Despondents, and SAR Training   4 hours
Outdoor First-Aid   8 & 16 hours
Fatigue and SAR Safety   3 hours

Incident Management Texts

☐ **Field Operations Guide for Search & Rescue:** Standard Operating Guidelines for Search & Rescue using the Incident Command System.
This book delivers concise easy-to-follow descriptions of ICS functions used in Search & Rescue. It goes beyond the normal descriptions found in ICS manuals and provides practical insight into the details necessary to successfully fulfill the SAR job. The writing and layout are clear and easy to follow, perfect for use in the field. For any SAR agency following ICS this book is a must read.
ISBN: 1-879471-15-9   1996 68p. (8" x 5.5")   **$9.95**

☐ **Incident Commander for Ground Search and Rescue**: A training course for Incident Commanders, Command Staff, Operations Section Chief, Planning Section Chief, base positions including the special topics of adaptation of the Incident Command System for SAR, managing information systems and protocols, ground OPS for aircraft, NTAP interpretation, specialized resource utilization and planning factors, high level legal implications, fatigue, extensive appendices, and more. Loaded with practical exercises. For anyone who has ever taken a MSO/MSF/MLPI course, this is the next step.
ISBN: 1-879471-21-3 1997 Instructors Manual   260 pages, 8½x11"   **$75.00**
ISBN: 1-879471-22-1 1997 Student's Manual   230 pages, 8½x5"   **$35.00**
To arrange classes please call.

☐ **Fatigue : Sleep Management During Disasters and Sustained Operations:** This training course allows the student to understand sleep physiology, required sleep, the use of naps, fatigue related accidents, and accident prevention. All searches and safety officers must be aware of what is often the most deadly aspect of search and rescue; the drive to and from the search!
ISBN: 1-879471-17-5 1997 Instructors Manual with overhead masters and MS-Powerpoint Showtime 3.5" disk, 8½x11"   **$110.00**
ISBN: 1-879471-18-3 1997 Student Manual 58 pages 5½x8'   **$9.95**
ISBN: 1-879471-19-1 1997 60 Color Overhead transparencies   **$200.00**
ISBN: 1-879471-20-5   1997 60 Slides   **$200.00**
To arrange seminars or consulting please call.

Wilderness Medicine

☐ **Outdoor First-Aid Training:** A wilderness First-Aid training course designed for trip leaders and search and rescue providers. Special emphasis on prevention and early recognition of disorders common in the outdoors. Topics include first-aid kits, dehydration, heat and cold disorders, bites and stings, injuries, and general medicine. Course is full of practical time and outdoor scenarios. Fundamentals version requires approximately 16 hours and the Essentials course takes approximately 8 hours.

ISBN: 1-879471-23-X Outdoor First-Aid Instructor's Manual 260 pages 8½ x11"	**$60.00**
ISBN: 1-879471-24-8 Essentials of Outdoor First-Aid 90 pages 8½x11"	**$10.00**
ISBN: 1-879471-25-6 Fundamentals of Outdoor First-Aid 110 pages 8½x11"	**$12.00**

For classes please call.

☐ **Outdoor First-Aid:** One of the few first-aid books that actually makes practical sense. 19 chapters with 57 illustrations and a good index and cross referencing. Covers well over 100 different common back country injuries and medical problems. Published on waterproof paper that floats! 1992

ISBN: 1-879471-14-0 Outdoor First-Aid 104 pages 4¼x7" appendix, index	**$13.95**

☐ **Wilderness and Rural Life Support Guidelines:** Ring bound in a pocket size on waterproof paper that floats. Snap rings bound to be customized with extensive appendixes or notes. WEMT level reference. Special emphasis on heat problems, hypothermia, altitude illnesses, venomous bites and stings. Appendix supplement includes: Suggested first-aid kit, water disinfection, poisonous plants & animals, signaling & mirror, field team leader check list, and track reporting form. 1991

ISBN: 1-879471-02-7 Plastic covered guidelines 60 pages, 4x7"	**$10.00**
Appendix supplement s-x	**$6.00**

☐ **Knots for Rescue:** If you have ever tried learning knots from a book and not quite gotten it, this video is for you. Shows practical applications in a typical, high-angle rescue scenario and then moves to provide detailed close ups with pause points between each knot.

ISBN: 1-879471-12-4 1990 50 minutes, full color, NTSC VHS	**$39.95**

Order Information

To reach us direct at dbS Productions LLC

Toll Free	(800) 745-1581
Fax	(434) 293-5502
Phone	(434) 293-5502
E-mail	robert@dbs-sar.com
Internet	www.dbs-sar.com